ENVIRONMENTAL REMEDIATION TEC[illegible]
REGULATIONS AND SAFETY

DEPLOYMENT OF CARBON CAPTURE AND SEQUESTRATION IN THE U.S.

BACKGROUND, DOE PROJECTS, AND FUTUREGEN 2.0

Environmental Remediation Technologies, Regulations and Safety

Additional books in this series can be found on Nova's website under the Series tab.

Additional e-books in this series can be found on Nova's website under the e-book tab.

ENVIRONMENTAL REMEDIATION TECHNOLOGIES, REGULATIONS AND SAFETY

DEPLOYMENT OF CARBON CAPTURE AND SEQUESTRATION IN THE U.S.

BACKGROUND, DOE PROJECTS, AND FUTUREGEN 2.0

DENIS TIERNAN
EDITOR

New York

For permission to use material from this book please contact us:
Telephone 631-231-7269; Fax 631-231-8175
Web Site: http://www.novapublishers.com

Library of Congress Cataloging-in-Publication Data

ISBN: 978-1-63117-726-2

Published by Nova Science Publishers, Inc. † New York

CONTENTS

PREFACE

Carbon capture and sequestration (or storage)—known as CCS—has attracted congressional interest as a measure for mitigating global climate change because large amounts of carbon dioxide (CO2) emitted from fossil fuel use in the United States are potentially available to be captured and stored underground and prevented from reaching the atmosphere. Large, industrial sources of CO2, such as electricity-generating plants, are likely initial candidates for CCS because they are predominantly stationary, single-point sources. Electricity generation contributes over 40% of U.S. CO2 emissions from fossil fuels. Currently, U.S. power plants do not capture large volumes of CO2 for CCS. This book covers only CCS and not other types of carbon sequestration activities, and it also aims to provide a snapshot of the DOE CCS program, including its current funding levels, together with some discussion of the program's achievements and prospects for success in meeting its stated goals.

Chapter 1 - Carbon capture and sequestration (or storage)—known as CCS—has attracted congressional interest as a measure for mitigating global climate change because large amounts of carbon dioxide (CO_2) emitted from fossil fuel use in the United States are potentially available to be captured and stored underground and prevented from reaching the atmosphere. Large, industrial sources of CO_2, such as electricity-generating plants, are likely initial candidates for CCS because they are predominantly stationary, single-point sources. Electricity generation contributes over 40% of U.S. CO_2 emissions from fossil fuels. Currently, U.S. power plants do not capture large volumes of CO_2 for CCS.

Several projects in the United States and abroad—typically associated with oil and gas production—are successfully capturing, injecting, and storing

CO_2 underground, albeit at relatively small scales. The oil and gas industry in the United States injects nearly 50 million tons of CO_2 underground each year for the purpose of enhanced oil recovery (EOR). The volume of CO_2 envisioned for CCS as a climate mitigation option is overwhelming compared to the amount of CO_2 used for EOR. According to the U.S. Department of Energy (DOE), the United States has the potential to store billions of tons of CO_2 underground and keep the gas trapped there indefinitely. Capturing and storing the equivalent of decades or even centuries of CO_2 emissions from power plants (at current levels of emissions) suggests that CCS has the potential to reduce U.S. greenhouse gas emissions substantially while allowing the continued use of fossil fuels.

An integrated CCS system would include three main steps: (1) capturing and separating CO_2 from other gases; (2) purifying, compressing, and transporting the captured CO_2 to the sequestration site; and (3) injecting the CO_2 in subsurface geological reservoirs or storing it in the oceans. Deploying CCS technology on a commercial scale would be a vast undertaking. The CCS process, although simple in concept, would require significant investments of capital and of time. Capital investment would be required for the technology to capture CO_2 and for the pipeline network to transport the captured CO_2 to the disposal site. Time would be required to assess the potential CO_2 storage reservoir, inject the captured CO_2, and monitor the injected plume to ensure against leaks to the atmosphere or to underground sources of drinking water, potentially for years or decades until injection activities cease and the injected plume stabilizes.

Three main types of geological formations in the United States are being considered for storing large amounts of CO2: oil and gas reservoirs, deep saline reservoirs, and unmineable coal seams. The deep ocean also has a huge potential to store carbon; however, direct injection of CO_2 into the deep ocean is controversial, and environmental concerns have forestalled planned experiments in the open ocean. Mineral carbonation—reacting minerals with a stream of concentrated CO_2 to form a solid carbonate—is well understood, but it is still an experimental process for storing large quantities of CO_2.

Large-scale CCS injection experiments are underway in the United States to test how different types of reservoirs perform during CO_2 injection of 1 million tons of CO_2 per year or more. Results from the experiments will undoubtedly be crucial to future permitting and site approval regulations.

Acceptance by the general public of large-scale deployment of CCS may be a significant challenge. Some of the large-scale injection tests could garner information about public acceptance, as citizens become familiar with the concept, process, and results of CO_2 injection tests in their local communities.

Chapter 2 - On September 20, 2013, the U.S. Environmental Protection Agency (EPA) re-proposed standards for carbon dioxide (CO_2) emissions from new fossil-fueled power plants. On January 8, 2014, EPA published the re-proposed rule in the *Federal Register*, triggering the start of a 60-day public comment period. The proposed rule places a new focus on whether the U.S. Department of Energy's (DOE's) CCS research, development, and demonstration (RD&D) program will achieve its vision of developing an advanced CCS technology portfolio ready by 2020 for large-scale CCS deployment.

As re-proposed, the standards would limit emissions of CO_2 to no more than 1,100 pounds per megawatt-hour (lbs/Mwh) of production from new coal-fired power plants and between 1,000 and 1,100 lbs/Mwh (depending on size of the plant) for new natural gas-fired plants. EPA proposed the standard under Section 111 of the Clean Air Act. According to EPA, new natural gas-fired stationary power plants should be able to meet the proposed standards without additional cost and the need for add-on control technology. However, new coal-fired plants only would be able to meet the standards by installing carbon capture and sequestration (CCS) technology. The re-proposed rule has sparked increased scrutiny of the future of CCS as a viable technology for reducing CO_2 emissions from coal-fired power plants.

Congress appropriated $3.4 billion from the American Recovery and Reinvestment Act (Recovery Act) for CCS RD&D at DOE's Office of Fossil Energy in addition to annual appropriations for CCS. The large influx of funding for industrial-scale CCS projects was intended to accelerate development and deployment of CCS in the United States. Since enactment of the Recovery Act, DOE has shifted its RD&D emphasis to the demonstration phase of carbon capture technology. To date, however, there are no commercial ventures in the United States that capture, transport, and inject industrial-scale quantities of CO_2 solely for the purposes of carbon sequestration.

The success of DOE CCS demonstration projects will likely influence the future outlook for widespread deployment of CCS technologies as a strategy for preventing large quantities of CO_2 from reaching the atmosphere while U.S. power plants continue to burn fossil fuels, mainly coal. One project, the Kemper County Facility, has received $270 million from DOE under its Clean Coal Power Initiative Round 2 program, and is slated to begin commercial operation in late 2014. The 583 megawatt capacity facility anticipates capturing 65% of its CO_2 emissions, making it equivalent to a new natural gas-fired combined cycle power plant. Cost overruns at the Kemper Plant, however, have raised questions over the relative value of environmental benefits due to CCS technology compared to construction costs of the facility and its effect on ratepayers.

Given the pending EPA rule, congressional interest in the future of coal as a domestic energy source appears directly linked to the future of CCS. Following the September 20, 2013, re-proposal of the rule the debate has been mixed as to whether the rule would spur development and deployment of CCS for new coal-fired power plants or have the opposite effect. Several bills introduced in the House and Senate, such as H.R. 3826 and S. 1905, directly address EPA's authority to issue regulations curtailing CO_2 emissions from coal-fired power plants. Congressional oversight of the CCS RD&D program could help inform decisions about the level of support for the program and help Congress gauge whether it is on track to meet its goals.

Chapter 3 - More than a decade after the George W. Bush Administration announced its signature clean coal power initiative—FutureGen—the program is still in early development. Since its inception in 2003, FutureGen has undergone changes in scope and design. As initially conceived, FutureGen would have been the world's first coal-fired power plant to integrate carbon capture and sequestration (CCS) with integrated gasification combined cycle (IGCC) technologies. FutureGen would have captured and stored carbon dioxide (CO_2) emissions from coal combustion in deep underground saline formations and produced hydrogen for electricity generation and fuel cell research. Increasing costs of development, among other considerations, caused the Bush Administration to discontinue the project in 2008. In 2010, under the Obama Administration, the project was restructured as FutureGen 2.0: a coal-fired power plant that would integrate oxycombustion technology to capture CO_2. FutureGen 2.0 is the U.S. Department of Energy's (DOE) most comprehensive CCS demonstration project, combining all three aspects of CCS technology: capturing and separating CO_2 from other gases, compressing and transporting CO_2 to the sequestration site, and injecting CO_2 in geologic formations for permanent storage.

Congressional interest in CCS technology centers on balancing the competing national interests of fostering low-cost, domestic sources of energy like coal against mitigating the effects of CO_2 emissions in the atmosphere. FutureGen 2.0 would address these interests by demonstrating CCS technology as commercially viable. Among the challenges to the development of FutureGen 2.0 are rising costs of production, ongoing issues with project development, lack of incentives for investment from the private sector, and time constraints. Further, FutureGen's development would need to include securing private sector funding to meet increasing costs, purchasing the power plant for the project, obtaining permission from DOE to retrofit the plant, performing the retrofit, and then meeting the goal of 90% capture of CO_2.

The FutureGen project was conceived as a public-private partnership between industry and DOE with agreements for cost-share and cooperation on development, demonstration, and deployment of CCS technology. The public-private partnership has been criticized for leading to setbacks in FutureGen's development, since the private sector lacks incentives to invest in costly CCS technology. Regulations, tax credits, or policies such as carbon taxation or cap-and-trade that increase the price of electricity from conventional power plants may be necessary to make CCS technology competitive enough for private sector investment. Even then, industry may choose to forgo coal-fired plants for other sources of energy that emit less CO_2, such as natural gas. However, Congress signaled its support for FutureGen 2.0 via the American Recovery and Reinvestment Act of 2009 (ARRA, P.L. 111-5) by appropriating almost $1 billion for the project. ARRA funding will expire on September 30, 2015, and it remains a question whether the project will expend all of its federal funding before that deadline.

A proposed rule by the Environmental Protection Agency (EPA) to limit CO_2 emissions from new fossil-fuel power plants may provide some incentive for industry to invest in CCS technology. The debate has been mixed as to whether the rule would spur development and deployment of CCS for new coal-fired power plants or have the opposite effect. Multiple analyses indicate that there will be retirements of U.S. coal-fired capacity; however, virtually all analyses agree that coal will continue to play a substantial role in electricity generation for decades. The rapid increase in the domestic natural gas supply as an alternative to coal, in combination with regulations that curtail CO_2 emissions, may lead electricity producers to invest in natural gas-fired plants, which emit approximately half the amount of CO_2 per unit of electricity produced compared to coal-fired plants.

In: Deployment of Carbon Capture and … ISBN: 978-1-63117-726-2
Editor: Denis Tiernan

Chapter 1

CARBON CAPTURE AND SEQUESTRATION (CCS): A PRIMER*

Peter Folger

SUMMARY

Carbon capture and sequestration (or storage)—known as CCS—has attracted congressional interest as a measure for mitigating global climate change because large amounts of carbon dioxide (CO_2) emitted from fossil fuel use in the United States are potentially available to be captured and stored underground and prevented from reaching the atmosphere. Large, industrial sources of CO_2, such as electricity-generating plants, are likely initial candidates for CCS because they are predominantly stationary, single-point sources. Electricity generation contributes over 40% of U.S. CO_2 emissions from fossil fuels. Currently, U.S. power plants do not capture large volumes of CO_2 for CCS.

Several projects in the United States and abroad—typically associated with oil and gas production—are successfully capturing, injecting, and storing CO_2 underground, albeit at relatively small scales. The oil and gas industry in the United States injects nearly 50 million tons of CO_2 underground each year for the purpose of enhanced oil

* This is an edited, reformatted and augmented version of a Congressional Research Service publication, CRS Report for Congress R42532, prepared for Members and Committees of Congress, from www.crs.gov, dated July 16, 2013.

recovery (EOR). The volume of CO_2 envisioned for CCS as a climate mitigation option is overwhelming compared to the amount of CO_2 used for EOR. According to the U.S. Department of Energy (DOE), the United States has the potential to store billions of tons of CO_2 underground and keep the gas trapped there indefinitely. Capturing and storing the equivalent of decades or even centuries of CO_2 emissions from power plants (at current levels of emissions) suggests that CCS has the potential to reduce U.S. greenhouse gas emissions substantially while allowing the continued use of fossil fuels.

An integrated CCS system would include three main steps: (1) capturing and separating CO_2 from other gases; (2) purifying, compressing, and transporting the captured CO_2 to the sequestration site; and (3) injecting the CO_2 in subsurface geological reservoirs or storing it in the oceans. Deploying CCS technology on a commercial scale would be a vast undertaking. The CCS process, although simple in concept, would require significant investments of capital and of time. Capital investment would be required for the technology to capture CO_2 and for the pipeline network to transport the captured CO_2 to the disposal site. Time would be required to assess the potential CO_2 storage reservoir, inject the captured CO_2, and monitor the injected plume to ensure against leaks to the atmosphere or to underground sources of drinking water, potentially for years or decades until injection activities cease and the injected plume stabilizes.

Three main types of geological formations in the United States are being considered for storing large amounts of CO2: oil and gas reservoirs, deep saline reservoirs, and unmineable coal seams. The deep ocean also has a huge potential to store carbon; however, direct injection of CO_2 into the deep ocean is controversial, and environmental concerns have forestalled planned experiments in the open ocean. Mineral carbonation—reacting minerals with a stream of concentrated CO_2 to form a solid carbonate—is well understood, but it is still an experimental process for storing large quantities of CO_2.

Large-scale CCS injection experiments are underway in the United States to test how different types of reservoirs perform during CO_2 injection of 1 million tons of CO_2 per year or more. Results from the experiments will undoubtedly be crucial to future permitting and site approval regulations.

Acceptance by the general public of large-scale deployment of CCS may be a significant challenge. Some of the large-scale injection tests could garner information about public acceptance, as citizens become familiar with the concept, process, and results of CO_2 injection tests in their local communities.

INTRODUCTION

Carbon capture and sequestration (or storage)—known as CCS—is a physical process that involves capturing manmade carbon dioxide (CO_2) at its source and storing it before its release to the atmosphere. CCS could reduce the amount of CO_2 emitted to the atmosphere despite the continued use of fossil fuels.

An integrated CCS system would include three main steps: (1) capturing CO_2 and separating it from other gases; (2) purifying, compressing, and transporting the captured CO_2 to the sequestration site; and (3) injecting the CO_2 in subsurface geological reservoirs or storing it in the oceans. As a measure for mitigating global climate change, CCS has attracted congressional interest and support because several projects in the United States and abroad—typically associated with oil and gas production—are successfully capturing, injecting, and storing CO_2 underground, albeit at relatively small scales. The oil and gas industry in the United States injects nearly 50 million metric tons of CO_2 underground each year to help recover oil and gas resources (a process known as enhanced oil recovery, or EOR).[1] Potentially, much larger amounts of CO_2 produced from electricity generation—approximately 2.1 billion metric tons per year, about 40% of the total CO_2 emitted in the United States from fossil fuels (see *Table 1*)—could be targeted for large-scale CCS.

Fuel combustion accounts for 94% of all U.S. CO_2 emissions.[2] Electricity generation contributes the largest proportion of CO_2 emissions compared to other types of fossil fuel use in the United States (*Table 1*). Electricity-generating plants are among the most likely initial candidates for capture, separation, and storage or reuse of CO_2 because they are predominantly large, stationary, single-point sources of emissions. Large industrial facilities, such as cement-manufacturing, ethanol, or hydrogen production plants, that produce large quantities of CO_2 as part of the industrial process are also good candidates for CO_2 capture and storage.[3]

This report is a brief summary of what CCS is, how it is supposed to work, why it has gained the interest and support of some Members of Congress, and what some of the challenges are to its implementation and deployment across the United States.

This report covers only CCS and not other types of carbon sequestration activities whereby CO_2 is removed from the atmosphere and stored in vegetation or soils, such as forests and agricultural lands.[4]

Table 1. Sources for CO_2 Emissions in the United States from Combustion of Fossil Fuels

Sources	CO_2 Emissions[a] (millions of metric tons)	Percent of Total
Electricity generation	2,108.8	41%
Transportation	1,745.0	34%
Industrial	773.2	15%
Residential	328.8	6%
Commercial	222.1	4%
Total	5,177.9	100%

Source: U.S. Environmental Protection Agency (EPA), Inventory of U.S. Greenhouse Emissions and Sinks: 1990- 2011, Table ES-3 (April 2013); http://www.epa.gov/climatechange/ghgemissions/usinventoryreport.html.

[a] CO_2 emissions in millions of metric tons for 2011; excludes emissions from U.S. territories.

CO_2 CAPTURE

The first step in CCS is to capture CO_2 at the source and produce a concentrated stream for transport and storage. Currently, three main approaches are available to capture CO_2 from large-scale industrial facilities or power plants: (1) post-combustion capture, (2) pre-combustion capture, and (3) oxy-fuel combustion capture. For power plants, current commercial CO_2 capture systems could operate at 85%-95% capture efficiency.[5] The capture phase of the CCS process, however, may be 80% or more of the total costs for CCS.[6]

Post-Combustion Capture

This process involves extracting CO_2 from the flue gas following combustion of fossil fuels or biomass. Several commercially available technologies, some involving absorption using chemical solvents, can in principle be used to capture large quantities of CO_2 from flue gases. U.S. commercial electricity-generating plants currently do not capture large volumes of CO_2 because they are not required to and there are no economic incentives to do so. Nevertheless, the post-combustion capture process includes proven technologies that are commercially available today.

Pre-Combustion Capture

This process separates CO_2 from the fuel by combining the fuel with air and/or steam to produce hydrogen for combustion and a separate CO_2 stream that could be stored. The most common technologies today use steam reforming, in which steam is employed to extract hydrogen from natural gas.[7]

Oxy-Fuel Combustion Capture

This process uses oxygen instead of air for combustion and produces a flue gas that is mostly CO_2 and water, which are easily separable, after which the CO_2 can be compressed, transported, and stored. The U.S. Department of Energy's (DOE's) flagship CCS demonstration project, FutureGen, plans to retrofit an existing power unit with an oxy-fuel combustion unit.[8]

CO_2 TRANSPORT

Pipelines are the most common method for transporting CO_2 in the United States. Currently, approximately 4,100 miles of pipeline transport CO_2 in the United States, predominately to oil and gas fields, where it is used for EOR.[9] Transporting CO_2 in pipelines is similar to transporting petroleum products like natural gas and oil; it requires attention to design, monitoring for leaks, and protection against overpressure, especially in populated areas.[10]

Using ships may be feasible when CO_2 must be transported over large distances or overseas. Ships transport CO_2 today, but at a small scale because of limited demand. Liquefied natural gas, propane, and butane are routinely shipped by marine tankers on a large scale worldwide. Rail cars and trucks can also transport CO_2, but this mode would probably be uneconomical for large-scale CCS operations.

Costs for pipeline transport vary, depending on construction, operation and maintenance, and other factors, including right-of-way costs, regulatory fees, and more. The quantity and distance transported will mostly determine costs, which will also depend on whether the pipeline is onshore or offshore, the level of congestion along the route, and whether mountains, large rivers, or frozen ground are encountered.

Shipping costs are unknown in any detail, however, because no large-scale CO_2 transport system (in millions of metric tons of CO_2 per year, for example) is operating. Ship costs might be lower than pipeline transport for distances greater than 1,000 kilometers and for less than a few million metric tons of CO_2 ($MtCO_2$)[11] transported per year.[12]

Even though regional CO_2 pipeline networks currently operate in the United States for enhanced EOR, developing a more expansive network for CCS could pose numerous regulatory and economic challenges. Some of these include questions about pipeline network requirements, economic regulation, utility cost recovery, regulatory classification of CO_2 itself, and pipeline safety.[13]

CO_2 SEQUESTRATION

Three main types of geological formations are being considered for carbon sequestration: (1) depleted oil and gas reservoirs, (2) deep saline reservoirs, and (3) unmineable coal seams. In each case, CO_2 would be injected in a supercritical state—a relatively dense liquid—below ground into a porous rock formation that holds or previously held fluids. When CO_2 is injected at depths greater than 800 meters in a typical reservoir, the pressure keeps the injected CO_2 in a supercritical state (dense like a liquid, fluid like a gas) and thus it is less likely to migrate out of the geological formation. Injecting CO_2 into deep geological formations uses existing technologies that have been primarily developed and used by the oil and gas industry, and that could potentially be adapted for long-term storage and monitoring of CO_2. Other underground injection applications in practice today, such as natural gas storage, deep injection of liquid wastes, and subsurface disposal of oil-field brines, can also provide valuable experience and information for sequestering CO_2 in geological formations.[14]

The storage capacity for CO_2 in geological formations is potentially huge if all the sedimentary basins in the world are considered (see discussion below of storage capacity estimates for the United States).[15] The suitability of any particular site, however, depends on many factors, including proximity to CO_2 sources and other reservoir-specific qualities like porosity, permeability, and potential for leakage.

For CCS to succeed, it is assumed that each reservoir type would permanently store the vast majority of injected CO_2, keeping the gas isolated from the atmosphere in perpetuity.

Oil and Gas Reservoirs

Pumping CO_2 into oil and gas reservoirs to boost production (EOR) is practiced in the petroleum industry today. The United States is a world leader in this technology, and oil and gas operators inject approximately 48 $MtCO_2$ underground each year to help recover oil and gas resources.[16] Most of the CO_2 used for EOR in the United States comes from naturally occurring geologic formations, however, not from industrial sources. Using CO_2 from industrial emitters has appeal because the costs of capture and transport from the facility could be partially offset by revenues from oil and gas production.

Carbon dioxide can be used for EOR onshore or offshore. To date, most CO_2 projects associated with EOR are onshore, with the bulk of U.S. activities in west Texas. Carbon dioxide can also be injected into oil and gas reservoirs that are completely depleted, which would serve the purpose of long-term sequestration, but without any offsetting benefit from oil and gas production.

Advantages and Disadvantages

Depleted or abandoned oil and gas fields, especially in the United States, are considered prime candidates for CO_2 storage for several reasons:

- oil and gas originally trapped did not escape for millions of years, demonstrating the structural integrity of the reservoir;
- extensive studies for oil and gas typically have characterized the geology of the reservoir;
- computer models have often been developed to understand how hydrocarbons move in the reservoir, and the models could be applied to predicting how CO_2 could move; and
- infrastructure and wells from oil and gas extraction may be in place and might be used for handling CO_2 storage.

Some of these features could also be disadvantages to CO_2 sequestration. Wells that penetrate from the surface to the reservoir could be conduits for CO_2 release if they are not plugged properly. Care must be taken not to overpressure the reservoir during CO_2 injection, which could fracture the caprock—the part of the formation that formed a seal to trap oil and gas—and subsequently allow CO_2 to escape. Also, shallow oil and gas fields (those less than 800 meters deep, for example) may be unsuitable because CO_2 may form a gas instead of a denser liquid and could escape to the surface more easily. In addition, oil and gas fields that are suitable for EOR may not necessarily be located near industrial sources of CO_2. Costs to construct pipelines to connect sources of CO_2 with oil and gas fields may, in part, determine whether an EOR operation using industrial sources of CO_2 is feasible.

Although the United States injects nearly 50 $MtCO_2$ underground each year for the purposes of EOR, that amount represents approximately 2% of the CO_2 emitted from fossil fuel electricity generation alone. The sheer volume of CO_2 envisioned for CCS as a climate mitigation option is overwhelming compared to the amount of CO_2 used for EOR. It may be that EOR will increase in the future, depending on economic, regulatory, and technical factors, and more CO_2 will be sequestered as a consequence. It is also likely that EOR would only account for a small fraction of the total amount of CO_2 injected underground in the future, even if CCS becomes a significant component in an overall scheme to substantially reduce CO_2 emissions to the atmosphere.

The In Salah and Weyburn Projects

The In Salah Project in Algeria was the world's first large-scale effort to store CO_2 in a natural gas reservoir.[17] At In Salah, CO_2 was separated from the produced natural gas (the gas contains approximately 5.5% CO_2) and then reinjected into the same formation. Approximately 17 $MtCO_2$ were planned to be captured and stored over the lifetime of the project at a rate of slightly more than 1 Mt per year.[18] Injection at In Salah ceased in 2011 and a future injection strategy is currently under review.

The Weyburn Project in south-central Canada uses CO_2 produced from a coal gasification plant in North Dakota for EOR, injecting up to 5,000 tCO_2 per day into the formation and enhancing oil recovery.[19] Approximately 20 $MtCO_2$ are expected to remain in the formation over the lifetime of the project.[20]

Deep Saline Reservoirs

Some rocks in sedimentary basins contain saline fluids—brines or brackish water unsuitable for agriculture or drinking. As with oil and gas, deep saline reservoirs can be found onshore and offshore; in fact, they are often part of oil and gas reservoirs and share many characteristics. The oil industry routinely injects brines recovered during oil production into saline reservoirs for disposal.[21]

Advantages and Disadvantages

Using suitably deep saline reservoirs for CO_2 sequestration has advantages: (1) they are more widespread in the United States than oil and gas reservoirs and thus have greater probability of being close to large point sources of CO_2; and (2) saline reservoirs have potentially the largest reservoir capacity of the three types of geologic formations.

Although deep saline reservoirs potentially have huge capacity to store CO_2, estimates of lower and upper capacities vary greatly, reflecting a higher degree of uncertainty in how to measure storage capacity.[22] Actual storage capacity may have to be determined on a case-by-case basis. Estimates of storage capacity for the United States from the DOE Regional Sequestration Partnership Program are discussed below.

From estimates of the potential storage capacity in saline reservoirs, it is likely that the vast majority of CO_2 injected underground would be stored in these formations, assuming that CCS were deployed on a commercial scale across the United States. In addition to their potential capacity, deep saline reservoirs underlie large portions of the country, and could be more easily accessible to large, stationary sources of CO_2 than oil and gas reservoirs or coal seams. *Figure 1* shows broad outlines of sedimentary basins containing the deep saline reservoirs, and the locations of a variety of stationary sources of CO_2.

The DOE Regional Sequestration Partnership Program has conducted simulations, field studies, and small-scale injection projects, and is now beginning a phase of large-scale injection demonstration projects to investigate the suitability of deep saline reservoirs.[23] Because of the potentially vast amounts of CO_2 that could be sequestered, these experiments could shed light on the potential for leakage of CO_2 from the reservoir, and test the ability to detect the movement of CO_2 underground as well as to detect leaks through overlying cap rocks.

In addition to the possibility of CO_2 leakage, injection of millions of tons of CO_2 will displace large volumes of brine in the deep saline reservoirs. One disadvantage is therefore the possibility that displaced brine could leak into underground sources of drinking water. Ultimately, CO_2 will likely dissolve into the brine, but that could take decades. Also, injecting large volumes of fluid into the subsurface has the potential to trigger earthquakes, especially if the CO_2 is injected into an undetected fault. Presumably, evaluating the potential storage site prior to beginning injection will limit the potential for triggering earthquakes (also referred to as "induced seismicity"), but there is no guarantee that fluid could not migrate to faulted or fractured rocks over the course of many years and induce an earthquake. The issue of induced seismicity has been linked to the injection and disposal of produced waters from oil and gas fields.[24]

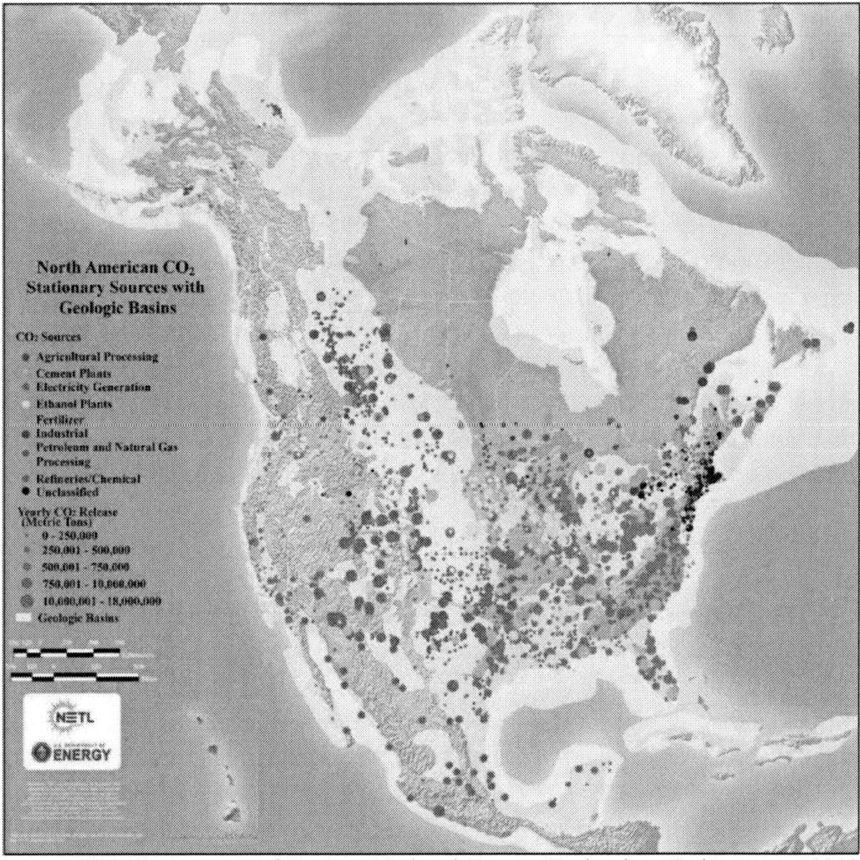

Source: U.S. Department of Energy, National Energy Technology Laboratory, "2010 Carbon Sequestration Atlas of the United States and Canada, Third Edition," http://www.netl.doe.gov/technologies/carbon_seq/refshelf/ atlasIII/index.html.

Note: Not all geologic basins have deep saline reservoirs suitable for carbon sequestration.

Figure 1. Stationary Sources of CO_2 in North America and Underlying Geologic Basins.

Unlike sequestration in existing oil and gas fields, injecting into deep saline reservoirs may take place in regions of the country that have not experienced drilling activities. Public opposition may arise to activities on the surface—such as land clearing, building of new roads, transport of heavy equipment, and operation of drill rigs—but also to the concept of disposing CO_2 underground near residences and communities. There is at least one example of public opposition to CO_2 injection leading to cancellation of a project in Europe.[25]

The Sleipner Project

The Sleipner Project in the North Sea is the first commercial-scale operation for sequestering CO_2 in a deep saline reservoir. The Sleipner project has been operating since 1996, and it injects and stores approximately 2,800 tCO_2 per day, or about 1 $MtCO_2$ per year.[26] Carbon dioxide is separated from natural gas production at the nearby Sleipner West Gas Field, compressed, and then injected 800 meters below the seabed of the North Sea into the Utsira formation, a sandstone reservoir 200-250 meters (650-820 feet) thick containing saline fluids. Monitoring has indicated the CO_2 has not leaked from the saline reservoir, and computer simulations suggest that the CO_2 will eventually dissolve into the saline water, reducing the potential for leakage in the future.

Another CO_2 sequestration project, similar to Sleipner, began in the Barents Sea in April 2008 (the Snohvit Project),[27] and is injecting approximately 2,000 tCO_2 per day below the seafloor. A larger project is being planned in western Australia (the Gorgon Project)[28] that would inject 9,000 tCO_2 per day when at full capacity (over 14 $MtCO_2$ per year). Similar to the Sleipner and Snohvit operations, the Gorgon plans to strip CO_2 from produced natural gas and inject it into deep saline formations for permanent storage. The Gorgon Project is scheduled to begin in 2015.

Unmineable Coal Seams

U.S. coal resources that are not mineable with current technology are those where the coal beds[29] are not thick enough, or are too deep, or whose structural integrity is inadequate for mining. Even if they cannot be mined, coal beds are commonly permeable and can trap gases, such as methane, which can be extracted (a resource known as coal-bed methane, or CBM). Methane and other gases are physically bound (adsorbed) to the coal. Studies indicate that CO_2 binds even more tightly to coal than methane.[30] Carbon dioxide injected into permeable coal seams could displace methane, which could be recovered by wells and brought to the surface, providing a source of revenue to offset the costs of CO_2 injection.

Advantages and Disadvantages

Unmineable coal seam injection projects would need to assess several factors in addition to the potential for CBM extraction. These include depth, permeability, coal bed geometry (a few thick seams, not several thin seams), lateral continuity and vertical isolation (less potential for upward leakage), and other considerations. Once CO_2 is injected into a coal seam, it would likely remain there unless the seam is depressurized or the coal is mined. Many unmineable coal seams in the United States are located relatively near electricity-generating facilities, which could reduce the distance and cost of transporting CO_2 from large point sources to storage sites.

Not all types of coal beds are suitable for CBM extraction. Without the coal-bed methane resource, the sequestration process would be less economically attractive. However, the displaced methane would need to be combusted or captured because methane itself is a more potent greenhouse gas than CO_2. Once burned, methane produces mostly CO_2 and water.

Without ongoing commercial experience, storing CO_2 in coal seams has significant uncertainties compared to the other two types of geological storage discussed. According to IPCC, unmineable coal seams have the smallest potential capacity for storing CO_2 globally compared to oil and gas fields or deep saline formations. The latest assessment from DOE also indicates that unmineable coal seams in the United States have less potential capacity than U.S. oil and gas fields for storing CO_2. (See following discussion.) No commercial CO_2 injection and sequestration projects in coal beds are currently underway in the United States.

Geological Storage Capacity for CO_2 in the United States

DOE National Technology Laboratory Carbon Sequestration Atlas

As *Figure 1* indicates, geologic basins containing at least one of each of these three types of potential CO_2 reservoirs occur across most of the United States, in relative proximity to many large point sources of CO_2, such as fossil fuel power plants or cement plants. The DOE Regional Sequestration Partnership Program has produced estimates of the potential storage capacity for each of these types of reservoirs and published the estimates in a Carbon Sequestration Atlas. The 2012 Carbon Sequestration Atlas (fourth edition) updates the 2010 version (third edition), and a summary of the storage estimates for both editions is compared in *Table 2*.[31]

The Carbon Sequestration Atlas was compiled from estimates of geological storage capacity made by seven separate regional partnerships (government-industry collaborations fostered by DOE) that each produced estimates for different regions of the United States and parts of Canada. According to DOE, the department determined early in the program that addressing CO_2 mitigation from power and industrial sources regionally would be the most effective way to address differences in geology, climate, population density, infrastructure, and socioeconomic development throughout the United States.[32] The Carbon Sequestration Atlas reflects some of the regional differences; for example, not all of the regional partnerships identified unmineable coal seams as potential CO_2 reservoirs. Other partnerships identified geological formations unique to their regions—such as organic-rich shales in the Illinois Basin, or flood basalts in the Columbia River Plateau—as other types of possible reservoirs for CO_2 storage.

Table 2 indicates a lower and upper range for sequestration potential in deep saline formations and for unmineable coal seams, but only a single estimate for oil and gas fields. Comparison between the 2010 and 2012 estimates indicates large changes between the two estimates for oil and gas fields and for the lower estimate for deep saline formations, but relatively small changes for the upper estimate for deep saline formations and unmineable coal seams. It is clear from the table that DOE considers estimates for oil and gas fields much better constrained by available data than for the other types of reservoirs. The amount and types of data from oil and gas fields, such as production history, and reservoir volume calculations, often represent decades of experience in the oil and gas industry. In the Carbon Sequestration Atlas, oil and gas reservoirs were assessed at the field level (i.e., on a finer scale and in more detail) than deep saline formations or unmineable coal seams, which were assessed at the basin level (i.e., at a coarser scale and in less detail).

A frequently asked question is how the estimates of storage capacity in the United States equate to years of storage, assuming a certain level of CO_2 emissions. The total lower estimate (sum of the three reservoir types) from the 2012 Carbon Sequestration Atlas shown in *Table 2* indicates the potential to store the equivalent of over 1,100 years of CO_2 emissions from electricity generation in the United States at current emission rates (2.1 billion tons per year). The total upper estimate indicates the potential for over 9,000 years of CO_2 emissions from electricity generation. These projections assume that all of the capacity could be utilized for storage, and would change if the CO_2 emissions rate changed.

The Sequestration Atlas is updated approximately every two years. Estimates of storage capacity change, often substantially, from update to update, as demonstrated in *Table 2*. The DOE Sequestration Atlas should probably be considered an evolving assessment of U.S. reservoir capacity for CO_2 storage.[33]

Table 2. Geological Sequestration Potential for the United States and Parts of Canada (billion metric tons of CO_2)

Reservoir type	Lower estimate (2010)	Lower estimate (2012)	% Change	Upper estimate (2010)	Upper estimate (2012)	% Change
Oil and gas fields	143	226	+58%	143	226	+58%
Deep saline formations	1,653	2,102	+27%	20,213	20,043	-0.8%
Unmineable coal seams	60	56	-7%	117	114	-3%
Totals	1,856	2,384	+28%	20,473	20,383	-0.04%

Source: 2010 and 2012 Carbon Sequestration Atlases.

U.S. Geological Survey National Assessment

The Energy Independence and Security Act of 2007 (EISA, P.L. 110-140) directed the Department of the Interior (DOI) to develop a single methodology for an assessment of the national potential for geologic storage of carbon dioxide. The U.S. Geological Survey (USGS) released an initial methodology in 2009. In response to external comments and reviews, the USGS revised its initial methodology in a 2010 report.[34] On June 26, 2013, the USGS released its assessment of geologic CO_2 storage resources across the United States, following its previously developed methodology.[35]

The USGS obtained a mean estimate of 3,000 billion metric tons of CO_2 storage capacity below onshore areas in the United States and underlying state waters (approximately 3.5 miles from shore for most states; for Texas and the Florida Gulf of Mexico coast, the border extends about 10.3 miles from shore). The value obtained by the USGS is about 25% greater than the lower estimate from the DOE Carbon Sequestration Atlas for 2012 (*Table 2*) and is about 15% of the upper estimate from the DOE study. The USGS study differs from the DOE estimate in several ways. The USGS estimate was a geology-based examination of all sedimentary basins in the United States and underlying state waters, and was based on the same, probabilistic methodology for all basins. In contrast, the DOE estimates were derived from each of the seven regional CCS sequestration partnerships and not a single, uniform study. The DOE estimate included portions of Canada and the USGS study did not. Also, the DOE estimate included potential sequestration capacity in coal seams; the USGS study did not include coal seams.

Of the eight regions[36] examined in the USGS study, the Coastal Plain region accounted for two-thirds (2,000 billion metric tons) of the total storage capacity estimated, of which approximately 1,800 billion metric tons of capacity is along the U.S. Gulf Coast.37 (Figure 2 shows the eight regions examined in the USGS study.) The region with the next largest capacity is Alaska with 270 million metric tons, mostly along the North Slope. It is unlikely that Alaska would host a significant amount of CO2 storage because the state lacks facilities that emit large volumes of CO2 compared to the conterminous United States.

DEEP OCEAN SEQUESTRATION

The world's oceans contain approximately 50 times the amount of carbon stored in the atmosphere and nearly 10 times the amount stored in plants and soils.[38] The oceans today take up—act as a net sink for—approximately 1.7 billion metric tons of CO_2 ($GtCO_2$) per year. About 45% of the CO_2 released from fossil fuel combustion and land use activities during the 1990s has remained in the atmosphere, while the remainder has been taken up by the oceans, vegetation, or soils on the land surface.[39] Without the ocean sink, atmospheric CO_2 concentration would be increasing more rapidly. Ultimately, the oceans could store more than 90% of all the carbon released to the atmosphere by human activities, but the process takes thousands of years.[40] The ocean's capacity to absorb atmospheric CO_2 may change, however, and possibly even decrease in the future.[41] Also, studies indicate that as more CO_2 enters the ocean from the atmosphere, the surface waters are becoming more acidic.[42]

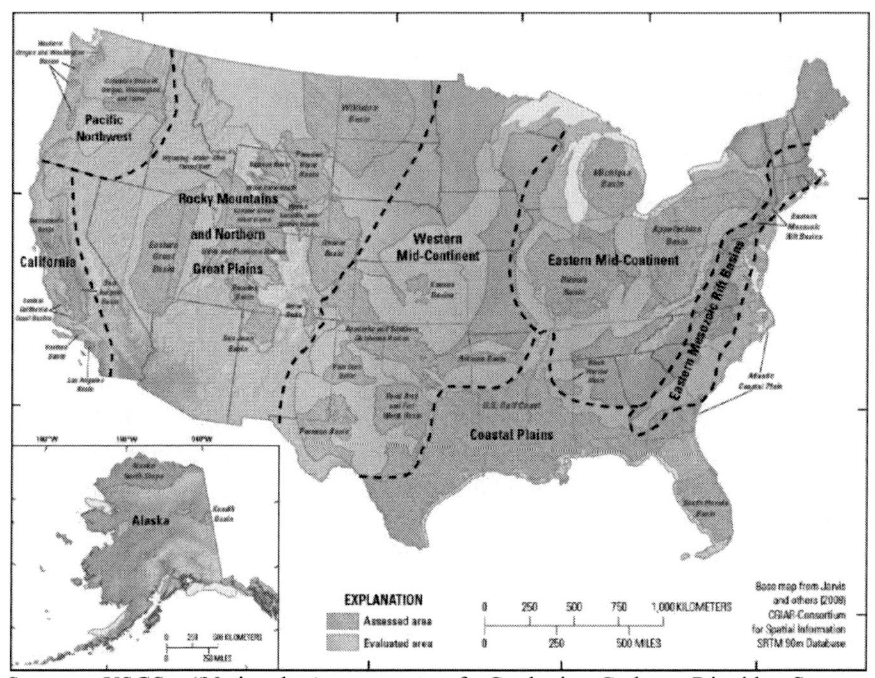

Source: USGS, "National Assessment of Geologic Carbon Dioxide Storage Resources—Results", USGS Circular 1386, 2013, http://pubs.usgs.gov/circ/1386/pdf/circular1386.pdf.

Notes: Assessed areas are shown by the pattern, evaluated areas are bluish gray and unpatterned. Evaluated areas were not assessed for their potential for CO2 storage. Resources in federally owned offshore areas were not assessed. Hawaii was not considered to have significant storage resources and was not assessed. Higher elevations are shown in brown and tan and lower elevations in green.

Figure 2. Map of the United States Showing Eight Regions Evaluated by the USGS for Potential CO_2 Storage.

Advantages and Disadvantages

Although the surface of the ocean is becoming more concentrated with CO_2, the surface waters and the deep ocean waters generally mix very slowly, on the order of decades to centuries. Injecting CO_2 directly into the deep ocean would take advantage of the slow rate of mixing, allowing the injected CO_2 to remain sequestered until the surface and deep waters mix and CO_2 concentrations equilibrate with the atmosphere. What happens to the CO_2 would depend on how it is released into the ocean, the depth of injection, and the temperature of the seawater.

Carbon dioxide injected at depths shallower than 500 meters typically would be released as a gas, and would rise towards the surface. Most of it would dissolve into seawater if the injected CO_2 gas bubbles were small enough.[43] At depths below 500 meters, CO_2 can exist as a liquid in the ocean, although it is less dense than seawater. After injection below 500 meters, CO_2 would also rise, but an estimated 90% would dissolve in the first 200 meters. Below 3,000 meters in depth, CO_2 is a liquid and is denser than seawater; the injected CO_2 would sink and dissolve in the water column or possibly form a CO_2 pool or lake on the sea bottom. Some researchers have proposed injecting CO_2 into the ocean bottom sediments below depths of 3,000 meters, and immobilizing the CO_2 as a dense liquid or solid CO_2 hydrate.[44] Deep storage in ocean bottom sediments, below 3,000 meters in depth, might potentially sequester CO_2 for thousands of years.[45]

The potential for ocean storage of captured CO_2 is huge, but environmental impacts on marine ecosystems and other issues may determine whether large quantities of captured CO_2 will ultimately be stored in the oceans. Also, deep ocean storage is in a research stage, and the effects of scaling up from small research experiments, using less than 100 liters of CO_2,[46] to injecting several $GtCO_2$ into the deep ocean are unknown.

Injecting CO_2 into the deep ocean would change ocean chemistry, locally at first, and assuming that hundreds of $GtCO_2$ were injected, would eventually produce measurable changes over the entire ocean.[47] The most significant and immediate effect would be the lowering of pH, increasing the acidity of the water. A lower pH may harm some ocean organisms, depending on the magnitude and rate of the pH change and the type of organism. Actual impacts of deep sea CO_2 sequestration are largely unknown, however, because scientists know very little about deep ocean ecosystems.[48]

Environmental concerns led to the cancellation of the largest planned experiment to test the feasibility of ocean sequestration in 2002. A scientific consortium had planned to inject 60 tCO_2 into water over 800 meters deep near the Kona coast on the island of Hawaii. Environmental organizations opposed the experiment on the grounds that it would acidify Hawaii's fishing grounds, and that it would divert attention from reducing greenhouse gas emissions.[49] A similar but smaller project with plans to release more than 5 tCO_2 into the deep ocean off the coast of Norway, also in 2002, was cancelled by the Norway Ministry of the Environment after opposition from environmental groups.[50]

Sequestering Under the Seabed

Deep ocean sequestration, as discussed here, is different from injecting CO_2 beneath the seabed into depleted oil and gas reservoirs or deep saline formations. The Sleipner project discussed above is an example of injection beneath the seafloor, but not injection into the ocean waters. Sequestering CO_2 under the seabed on the U.S. continental shelf would eliminate the need to negotiate with local landowners over the rights to surface land and to the pore space in the subsurface. However, it would also require developing an offshore infrastructure to transport and inject the captured CO_2, along with all the other challenges of evaluating the potential offshore reservoir, including monitoring the injected CO_2, and providing for liability and ownership of the CO_2 after injection has ceased.

Mineral Carbonation

Another option for sequestering CO_2 produced by fossil fuel combustion involves converting CO_2 to solid inorganic carbonates, such as $CaCO_3$ (limestone), using chemical reactions. When this process occurs naturally, it is known as "weathering" and takes place over thousands or millions of years. The process can be accelerated by reacting a high concentration of CO_2 with minerals found in large quantities on the Earth's surface, such as olivine or serpentine.[51] Mineral carbonation has the advantage of sequestering carbon in solid, stable minerals that can be stored without risk of releasing carbon to the atmosphere over geologic time scales.[52]

Mineral carbonation involves three major activities: (1) preparing the reactant minerals—mining, crushing, and milling—and transporting them to a processing plant, (2) reacting the concentrated CO_2 stream with the prepared minerals, and (3) separating the carbonate products and storing them in a suitable repository.

Advantages and Disadvantages

Mineral carbonation is well understood and can be applied at small scales, but is at an early phase of development as a technique for sequestering large amounts of captured CO_2. Large volumes of silicate oxide minerals are needed, from 1.6 to 3.7 metric tons of silicates per tCO_2 sequestered. Thus, a large-scale mineral carbonation process needs a large mining operation to provide the reactant minerals in sufficient quantity.[53] Large volumes of solid material would also be produced, between 2.6 and 4.7 metric tons of materials per tCO_2 sequestered, or 50%-100% more material to be disposed of by volume than originally mined. Because mineral carbonation is in the research and experimental stage, estimating the amount of CO_2 that could be sequestered by this technique is difficult.

One possible type of geological reservoir for CO_2 storage—major flood basalts[54] such as those on the Columbia River Plateau—is being explored for its potential to react with CO_2 and form solid carbonates in situ (in place). Instead of mining, crushing, and milling the reactant minerals, as discussed above, CO_2 would be injected directly into the basalt formations and would react with the rock over time and at depth to theoretically form solid carbonate minerals. Large and thick formations of flood basalts occur globally, and many have characteristics—such as high porosity and permeability—that are favorable to storing CO_2. Those characteristics, combined with the tendency of basalt to react with CO_2, could result in a large-scale conversion of the gas into stable, solid minerals that would remain underground for geologic time. The DOE regional carbon sequestration partnerships are exploring the possibility of using Columbia River Plateau flood basalts in the Pacific Northwest for storing CO_2.[55]

CURRENT ISSUES AND FUTURE CHALLENGES

A primary goal of developing and deploying CCS is to allow large industrial facilities, such as fossil fuel power plants and cement plants, to operate while reducing their CO_2 emissions by 80%-90%. Such reductions would presumably reduce the likelihood of continued climate warming from greenhouse gases by slowing the rise in atmospheric concentrations of CO_2. To achieve the overarching goal of reducing the likelihood of continued climate warming would depend, in part, on how fast and how widely CCS could be deployed throughout the economy.

The additional cost of installing CCS on CO_2-emitting facilities is a primary challenge to the adoption and deployment of CCS in the United

States, especially in an environment of low natural gas prices and increasing domestic natural gas supplies. Major increases in CO_2 capture technology efficiency will likely produce the greatest relative cost savings for CCS systems, but challenges also face the transport and storage components of CCS.

Ideally, storage reservoirs for CO_2 would be located close to sources, obviating the need to build a large pipeline infrastructure to deliver captured CO_2 for underground sequestration. If CCS moves to widespread implementation, however, some areas of the country may not have adequate reservoir capacity nearby, and may need to construct pipelines from sources to reservoirs. Identifying and validating sequestration sites would need to account for CO_2 pipeline costs, for example, if the economics of the sites are to be fully understood. If this is the case, there would be questions to be resolved regarding pipeline network requirements, economic regulation, utility cost recovery, regulatory classification of CO_2 itself, and pipeline safety. In addition, Congress may be called upon to address federal jurisdictional authority over CO_2 pipelines under existing law, and whether additional legislation may be necessary if a CO_2 pipeline network grows and crosses state lines.

Although DOE has identified substantial potential storage capacity for CO_2, particularly in deep saline formations, large-scale injection experiments are only beginning in the United States to test how different types of reservoirs perform during CO_2 injection. Data from the experiments will undoubtedly be crucial to future permitting and site approval regulations.

In addition, liability, ownership, and long-term stewardship for CO_2 sequestered underground are issues that would need to be resolved before CCS is deployed commercially. Some states are moving ahead with state-level geological sequestration regulations for CO_2, so federal efforts to resolve these issues at a national level would likely involve negotiations with the states. Acceptance by the general public of large-scale deployment of CCS may be a significant challenge if the majority of CCS projects involve private land. Some of the large-scale injection tests could garner information about public acceptance, as local communities become familiar with the concept, process, and results of CO_2 injection tests. Apart from the question of how the public would accept the likely higher cost for electricity generated from plants with CCS, how a growing CCS infrastructure of pipelines, injection wells, underground reservoirs, and other facilities would be accepted by the public is as yet unknown.

End Notes

[1] U.S. Department of Energy, National Energy Technology Laboratory, Carbon Sequestration Through Enhanced Oil Recovery, (March, 2008), at http://www.netl.doe.gov/publications/ factsheets/program/Prog053.pdf.

[2] U.S. Environmental Protection Agency (EPA), Inventory of U.S. Greenhouse Emissions and Sinks: 1990-2011, Table ES-2. The percentage refers to U.S. emissions in 2011; http://www.epa.gov/climatechange/Downloads/ghgemissions/ US-GHG-Inventory-2013-ES.pdf.

[3] Intergovernmental Panel on Climate Change (IPCC) Special Report: Carbon Dioxide Capture and Storage, 2005. (Hereafter referred to as IPCC Special Report.)

[4] For more information about carbon sequestration in forests and agricultural lands, see CRS Report RL31432, Carbon Sequestration in Forests, by Ross W. Gorte; CRS Report RL33898, Climate Change: The Role of the U.S. Agriculture Sector, by Renée Johnson; and CRS Report R40186, Biochar: Examination of an Emerging Concept to Sequester Carbon, by Kelsi Bracmort. For more information about carbon exchanges between the oceans, atmosphere, and land surface, see CRS Report RL34059, The Carbon Cycle: Implications for Climate Change and Congress, by Peter Folger.

[5] IPCC Special Report, p. 107.

[6] See, for example, John Deutch et al., The Future of Coal, Massachusetts Institute of Technology, An Interdisciplinary MIT Study, 2007, Executive Summary, p. xi.

[7] See CRS Report R41325, Carbon Capture: A Technology Assessment, by Peter Folger.

[8] See CRS Report R43028, FutureGen: A Brief History and Issues for Congress, by Peter Folger, for further discussion of FutureGen.

[9] Kevin Bliss et al., "A Policy, Legal, and Regulatory Evaluation of the Feasibility of a National Pipeline Infrastructure for the Transport and Storage of Carbon Dioxide," Interstate Oil and Gas Compact Commission, September 10, 2010, Table 3, http://www.sseb. org/downloads/pipeline.pdf. By comparison, nearly 500,000 miles of pipeline operate to convey natural gas and hazardous liquids in the United States.

[10] IPCC Special Report, p. 181.

[11] One metric ton of CO2 equivalent is written as 1 tCO2; one million metric tons is written as 1 MtCO2; one billion metric tons is written as 1 GtCO2.

[12] IPCC Special Report, p. 31.

[13] These issues are discussed in more detail in CRS Report RL33971, Carbon Dioxide (CO2) Pipelines for Carbon Sequestration: Emerging Policy Issues, by Paul W. Parfomak, Peter Folger, and Adam Vann, and CRS Report RL34316, Pipelines for Carbon Dioxide (CO2) Control: Network Needs and Cost Uncertainties, by Paul W. Parfomak and Peter Folger.

[14] IPCC Special Report, p. 31.

[15] Sedimentary basins refer to natural large-scale depressions in the Earth's surface that are filled with sediments and fluids and are therefore potential reservoirs for CO2 storage.

[16] Data from 2006. See DOE, National Energy Technology Laboratory, Carbon Sequestration Through Enhanced Oil Recovery, (March 2008), at http://www.netl.doe.gov/publications/ factsheets/program/Prog053.pdf.

[17] IPCC Special Report, p. 203.

[18] Injection started in 2004 and was suspended in 2011 due to concerns over the integrity of the geological seal that keeps the CO2 from leaking to the surface. The Carbon Capture and Sequestration Technologies Program at MIT, Carbon Capture and Sequestration Project Database, In Salah Fact Sheet, http://sequestration.mit.edu/tools/projects/ in_salah.html.

[19] IPCC Special Report, p. 204.

[20] MIT Carbon Capture and Sequestration Project Database, Weyburn Fact Sheet, http://sequestration.mit.edu/tools/ projects/weyburn.html.

[21] DOE Office of Fossil Energy; see http://www.fossil.energy.gov/programs/sequestration/ geologic/index.html.

[22] IPCC Special Report, p. 223.

[23] See CRS Report R42496, Carbon Capture and Sequestration: Research, Development, and Demonstration at the U.S. Department of Energy, by Peter Folger, for more information on the DOE programs.

[24] See, for example, Mike Soraghan, "Drilling Waste Disposal Risks Another Damaging Okla. Quake, Scientist Warns," Energywire, April 19, 2012, http://www.eenews.net/energywire/ 2012/04/19/archive/1?terms=earthquake.

[25] See Paul Voosen, "Public Outcry Scuttles German Demonstration Plant," Greenwire, December 6, 2011, http://www.eenews.net/Greenwire/2011/12/06/ archive/10?terms= vattenfall.

[26] Carbon Capture and Sequestration Project Database, Sleipner Fact Sheet, http://sequestration. mit.edu/tools/projects/ sleipner.html.

[27] Carbon Capture and Sequestration Project Database, Snohvit Fact Sheet, http://sequestration. mit.edu/tools/projects/ snohvit.html.

[28] Carbon Capture and Sequestration Project Database, Gorgon Fact Sheet, http://sequestration. mit.edu/tools/projects/ gorgon.html.

[29] Coal bed and coal seam are interchangeable terms.

[30] IPCC Special Report, p. 217.

[31] U.S. Dept. of Energy, National Energy Technology Laboratory, 2010 Carbon Sequestration Atlas of the United States and Canada, 3rd ed. (November 2010), 160 pages. Hereinafter referred to as the 2010 Carbon Sequestration Atlas, http://www.netl.doe.gov/technologies/ carbon_seq/refshelf/atlasIII/2010atlasIII.pdf; and 2012 Carbon Utilization and Storage Atlas—4th ed. (December 2012), 129 pages. Hereinafter referred to as the 2012 Carbon Sequestration Atlas, http://www.netl.doe.gov/technologies/carbon_seq/refshelf/atlas IV/Atlas-IV-2012.pdf.

[32] 2012 Carbon Sequestration Atlas, p. 7.

[33] 2010 Carbon Sequestration Atlas, p. 139.

[34] Sean T. Brennan et al., A Probabilistic Assessment Methodology for the Evaluation of Geologic Carbon Dioxide Storage, USGS, Open-File Report 2010-1127, 2010.

[35] U.S. Geological Survey, "National Assessment of Geologic Carbon Dioxide Storage Resources—Data," June 2012, http://pubs.usgs.gov/ds/774/.

[36] These include the Pacific Northwest, California, Rocky Mountains and Northern Great Plains, Western Mid-Continent, Eastern Mid-Continent, Eastern Mesozoic Rift Basins, Coastal Plains, and Alaska.

[37] USGS Fact Sheet 2013-3020, "National Assessment of Geologic Carbon Dioxide Storage Resources—Summary," http://pubs.usgs.gov/fs/2013/3020/.

[38] Christopher L. Sabine et al., "Current Status and Past Trends of the Global Carbon Cycle," in C. B. Field and M. R. Raupach, eds., The Global Carbon Cycle: Integrating Humans, Climate, and the Natural World (Washington, DC: Island Press, 2004), pp. 17-44.

[39] 2007 IPCC Working Group I Report, pp. 514-515.

[40] CO2 forms carbonic acid when dissolved in water. Over time, the solid calcium carbonate (CaCO3) on the seafloor will react with (neutralize) much of the carbonic acid that entered the oceans as CO2 from the atmosphere. See David Archer et al., "Dynamics of Fossil Fuel

CO_2 Neutralization by Marine $CaCO_3$," Global Biogeochemical Cycles, vol. 12, no. 2 (June 1998): pp. 259-276.

[41] One study, for example, suggests that the efficiency of the ocean sink has been declining at least since 2000; see Josep G. Canadell et al., "Contributions to Accelerating Atmospheric CO_2 Growth from Economic Activity, Carbon Intensity, and Efficiency of Natural Sinks," Proceedings of the National Academy of Sciences, vol. 104, no. 47 (Nov. 20, 2007), pp. 18866-18870.

[42] For more information on ocean acidification, see CRS Report R40143, Ocean Acidification, by Harold F. Upton and Peter Folger.

[43] IPCC Special Report, p. 285.

[44] A CO_2 hydrate is a crystalline compound formed at high pressures and low temperatures by trapping CO_2 molecules in a cage of water molecules.

[45] K. Z. House, et al., "Permanent Carbon Dioxide Storage in Deep-Sea Sediments," Proceedings of the National Academy of Sciences, vol. 103, no. 33 (Aug. 15, 2006): pp. 12291-12295.

[46] P. G. Brewer, et al., "Deep Ocean Experiments with Fossil Fuel Carbon Dioxide: Creation and Sensing of a Controlled Plume at 4 km Depth," Journal of Marine Research, vol. 63, no. 1 (2005): p. 9-33.

[47] IPCC Special Report, p. 279.

[48] Ibid., p. 298.

[49] Virginia Gewin, "Ocean Carbon Study to Quit Hawaii," Nature, vol. 417 (June 27, 2002): p. 888.

[50] Jim Giles, "Norway Sinks Ocean Carbon Study," Nature, vol. 419 (Sept. 5, 2002): p. 6.

[51] Serpentine and olivine are silicate oxide minerals—combinations of the silica, oxygen, and magnesium—that react with CO_2 to form magnesium carbonates. Wollastonite, a silica oxide mineral containing calcium, reacts with CO_2 to form calcium carbonate (limestone). Magnesium and calcium carbonates are stable minerals over long time scales.

[52] Calera, a company based in Los Gatos, CA, has developed a process for mineral carbonation that it claims will sequester CO_2 and produce solid carbonate minerals that can be used in the manufacture of building materials. The Calera process is discussed in a CRS congressional distribution (CD) memorandum, available from Peter Folger at 7- 1517.

[53] IPCC Special Report, p. 40.

[54] Flood basalts are vast expanses of solidified lava, commonly containing olivine, that erupted over large regions in several locations around the globe. In addition to the Columbia River Plateau flood basalts, other well-known flood basalts include the Deccan Traps in India and the Siberian Traps in Russia.

[55] 2010 Carbon Sequestration Atlas, p. 30.

In: Deployment of Carbon Capture and … ISBN: 978-1-63117-726-2
Editor: Denis Tiernan

Chapter 2

CARBON CAPTURE AND SEQUESTRATION: RESEARCH, DEVELOPMENT, AND DEMONSTRATION AT THE U.S. DEPARTMENT OF ENERGY*

Peter Folger

SUMMARY

On September 20, 2013, the U.S. Environmental Protection Agency (EPA) re-proposed standards for carbon dioxide (CO_2) emissions from new fossil-fueled power plants. On January 8, 2014, EPA published the re-proposed rule in the *Federal Register*, triggering the start of a 60-day public comment period. The proposed rule places a new focus on whether the U.S. Department of Energy's (DOE's) CCS research, development, and demonstration (RD&D) program will achieve its vision of developing an advanced CCS technology portfolio ready by 2020 for large-scale CCS deployment.

As re-proposed, the standards would limit emissions of CO_2 to no more than 1,100 pounds per megawatt-hour (lbs/Mwh) of production from new coal-fired power plants and between 1,000 and 1,100 lbs/Mwh (depending on size of the plant) for new natural gas-fired plants. EPA proposed the standard under Section 111 of the Clean Air Act. According

* This is an edited, reformatted and augmented version of Congressional Research Service Publication, No. R42496, dated February 10, 2014.

to EPA, new natural gas-fired stationary power plants should be able to meet the proposed standards without additional cost and the need for add-on control technology. However, new coal-fired plants only would be able to meet the standards by installing carbon capture and sequestration (CCS) technology. The re-proposed rule has sparked increased scrutiny of the future of CCS as a viable technology for reducing CO_2 emissions from coal-fired power plants.

Congress appropriated $3.4 billion from the American Recovery and Reinvestment Act (Recovery Act) for CCS RD&D at DOE's Office of Fossil Energy in addition to annual appropriations for CCS. The large influx of funding for industrial-scale CCS projects was intended to accelerate development and deployment of CCS in the United States. Since enactment of the Recovery Act, DOE has shifted its RD&D emphasis to the demonstration phase of carbon capture technology. To date, however, there are no commercial ventures in the United States that capture, transport, and inject industrial-scale quantities of CO_2 solely for the purposes of carbon sequestration.

The success of DOE CCS demonstration projects will likely influence the future outlook for widespread deployment of CCS technologies as a strategy for preventing large quantities of CO_2 from reaching the atmosphere while U.S. power plants continue to burn fossil fuels, mainly coal. One project, the Kemper County Facility, has received $270 million from DOE under its Clean Coal Power Initiative Round 2 program, and is slated to begin commercial operation in late 2014. The 583 megawatt capacity facility anticipates capturing 65% of its CO_2 emissions, making it equivalent to a new natural gas-fired combined cycle power plant. Cost overruns at the Kemper Plant, however, have raised questions over the relative value of environmental benefits due to CCS technology compared to construction costs of the facility and its effect on ratepayers.

Given the pending EPA rule, congressional interest in the future of coal as a domestic energy source appears directly linked to the future of CCS. Following the September 20, 2013, re-proposal of the rule the debate has been mixed as to whether the rule would spur development and deployment of CCS for new coal-fired power plants or have the opposite effect. Several bills introduced in the House and Senate, such as H.R. 3826 and S. 1905, directly address EPA's authority to issue regulations curtailing CO_2 emissions from coal-fired power plants. Congressional oversight of the CCS RD&D program could help inform decisions about the level of support for the program and help Congress gauge whether it is on track to meet its goals.

INTRODUCTION

Carbon capture and sequestration (or storage)—known as CCS—is a physical process that involves capturing manmade carbon dioxide (CO2) at its source and storing it before its release to the atmosphere. CCS could reduce the amount of CO2 emitted to the atmosphere while allowing the continued use of fossil fuels at power plants and other large, industrial facilities. An integrated CCS system would include three main steps: (1) capturing CO2 at its source and separating it from other gases; (2) purifying, compressing, and transporting the captured CO2 to the sequestration site; and (3) injecting the CO2 into subsurface geological reservoirs. Following its injection into a subsurface reservoir, the CO2 would need to be monitored for leakage and to verify that it remains in the target geological reservoir. Once injection operations cease, a responsible party would need to take title to the injected CO2 and ensure that it stays underground in perpetuity.

The U.S. Department of Energy (DOE) has pursued research and development of aspects of the three main steps leading to an integrated CCS system since 1997.[1] Congress has appropriated approximately $6 billion in total since FY2008 for CCS research, development, and demonstration (RD&D) at DOE's Office of Fossil Energy: approximately $3 billion in total annual appropriations (including FY2014), and $3.4 billion from the American Recovery and Reinvestment Act (P.L. 111-5, enacted February 17, 2009, hereinafter referred to as the Recovery Act).

The large and rapid influx of funding for industrial-scale CCS projects from the Recovery Act was intended to accelerate development and demonstration of CCS in the United States. The Recovery Act funding also was likely intended to help DOE achieve its RD&D goals as outlined in the department's 2010 RD&D *CCS Roadmap*.[2] (In part, the roadmap was intended to lay out a path for rapid technological development of CCS so that the United States would continue to use fossil fuels.) However, the future deployment of CCS may take a different course if the major components of the DOE program follow a path similar to DOE's FutureGen project, which has experienced delays and multiple changes of scope and design since its inception in 2003.[3]

This report aims to provide a snapshot of the DOE CCS program, including its current funding levels, together with some discussion of the program's achievements and prospects for success in meeting its stated goals. Other CRS reports provide substantial detail on the technological and policy aspects of CCS.[4]

EPA PROPOSED RULE: LIMITING CO2 EMISSIONS FROM POWER PLANTS[5]

New Power Plants

On September 20, 2013, the U.S. Environmental Protection Agency (EPA) re-proposed a standard that would limit emissions of carbon dioxide (CO2) from new fossil-fueled power plants. As re-proposed, the rule would limit emissions to no more than 1,100 pounds per megawatt-hour of electric generation from new coal-fired power plants and between 1,000 and 1,100 pounds per megawatt-hour (depending on size of the plant) for new natural gas-fired plants. EPA proposed the standard under Section 111 of the Clean Air Act. According to EPA, new natural gas-fired stationary power plants should be able to meet the proposed standard without additional cost and the need for add-on control technology. However, the only technical way for new coal-fired plants to meet the standard would be to install carbon capture and sequestration (CCS) technology to capture about 40% of the CO2 they typically produce. The proposed standard allows for a seven-year compliance period for coal-fired plants but would demand a more stringent standard for those plants that comply over seven years; CO2 emissions for these plants would be limited to an average of 1,000-1,050 pounds per megawatt-hour.[6]

On January 8, 2014, EPA published the re-proposed rule in the *Federal Register*.[7] Publishing in the *Federal Register* triggers the start of a 60-day public comment period: Comments will be accepted until March 10, 2014. The 2012 proposal generated more than 2.5 million comments, which prompted, in part, the September 20, 2013, re-proposal. Promulgation of the final rule could be expected sometime after the public comment period ends and EPA evaluates the comments.

The prospects for building new coal-fired electricity generating plants depend on many factors, such as costs of competing fuel sources (e.g., natural gas), electricity demand, regulatory costs, infrastructure (including rail) and electric grid development, and others. However, the EPA proposed rule clearly identifies CCS as the essential technology required if new coal-fired power plants are to be built in the United States.[8] The re-proposed standard places a new focus on DOE's CCS RD&D program—whether it will achieve its vision of "having an advanced CCS technology portfolio ready by 2020 for large-scale CCS demonstration that provides for the safe, cost-effective carbon management that will meet our Nation's goals for reducing [greenhouse gas] emissions."[9]

Existing Power Plants

The September 2013 re-proposed rule would address only new power plants. However, Section 111 of the Clean Air Act requires that EPA develop guidelines for greenhouse gas emissions for existing plants whenever it promulgates standards for new power plants. In his June 25, 2013, memorandum, President Obama directed the EPA to issue proposed guidelines for existing plants by June 1, 2014, and to issue final guidelines a year later.[10]

Implications for CCS Research, Development, and Deployment

Congress has appropriated funding for DOE to pursue CCS research and development since 1997 and signaled its interest in CCS technology by awarding $3.4 billion from the Recovery Act to CCS programs at DOE. Given the pending EPA rule, congressional interest in the future of coal as a domestic energy source appears directly linked to the future of CCS. Following the September 20, 2013, re-proposal of the rule, the debate has been mixed as to whether the rule would spur development and deployment of CCS for new coal-fired power plants or have the opposite effect. Multiple analyses indicate that there will be retirements of U.S. coal-fired capacity; however, virtually all analyses agree that coal will continue to play a substantial role in electricity generation for decades. How many retirements would take place and the role of EPA regulations in causing them are matters of dispute.[11]

Since the September 2013 re-proposal, the argument over the rule has focused, in part, on whether CCS is the best system of emissions reduction (BSER) for coal plants and whether it has been "adequately demonstrated" as such as required under the Clean Air Act. In its re-proposed rule, EPA cites the "existence and apparent ongoing viability" of several ongoing CCS demonstration projects as examples that justify a separate determination of BSER for coal-fired plants and integrated gasification combined-cycle plants. (The second BSER determination is for gas-fired power plants.)[12] The EPA noted that these projects had reached advanced stages of construction and development, "which suggests that proposing a separate standard for coal-fired units is appropriate."

The Natural Gas Alternative?

The huge increases in the U.S. domestic supply of natural gas in recent years, due largely to the exploitation of unconventional shale gas reservoirs through the use of hydraulic fracturing, has also led to a shift to natural gas for electricity production.[13] The shift appears to be largely due to the cheaper and increasingly abundant fuel—natural gas—compared to coal for electricity production. The EPA re-proposed rule, discussed above, noted that "power companies often choose the lowest cost form of generation when determining what type of new generation to build. Based on [Energy Information Administration] modeling and utility [Integrated Resource Plans], there appears to be a general acceptance that the lowest cost form of new power generations is [natural gas combined-cycle]." Cheap gas, due to the rapid increase in the domestic natural gas supply as an alternative to coal, in combination with regulations that curtail CO_2 emissions may lead electric power producers to invest in natural gas-fired plants, which emit approximately half the amount of CO_2 per unit of electricity produced compared to coal-fired plants. Regulations and abundant cheap gas may raise questions about the rationale for funding CCS demonstration projects like FutureGen.

Alternatively, and despite increasingly abundant domestic natural gas supplies, EPA regulations could provide the necessary incentives for the industry to accelerate CCS development and deployment for coal-fired power plants. As part of its re-proposed ruling, EPA cites technology as one of four factors that it considers in making a BSER determination.[14] Specifically, EPA stated that it "considers whether the system promotes the implementation and further development of technology," in this case referring to CCS technology. It appears that EPA asserts that its rule would likely promote CCS development and deployment rather than hinder it. Those arguing against the re-proposed rule do so on the basis that CCS technology has not been adequately demonstrated, and that it violates provisions in P.L. 109-58, the Energy Policy Act of 2005, that prohibit EPA from setting a performance standard based on the use of technology from certain DOE-funded projects, such as the three projects cited in the EPA re-proposal, among other reasons.[15]

LEGISLATION

Although DOE has pursued aspects of CCS RD&D since 1997, the Energy Policy Act of 2005 (P.L. 109-58) provided a 10-year authorization for the basic framework of CCS research and development at DOE.[16] The Energy Independence and Security Act of 2007 (EISA, P.L. 110-140) amended the Energy Policy Act of 2005 to include, among other provisions, authorization for seven large-scale CCS demonstration projects (in addition to FutureGen) that would integrate the carbon capture, transportation, and sequestration steps.[17] (Large-scale demonstration programs and their potential significance are discussed below.) It can be argued that, since enactment of EISA, the focus and funding within the CCS RD&D program has shifted toward large-scale capture technology development through these and other demonstration projects.

In addition to the annual appropriations provided for CCS RD&D, the Recovery Act (P.L. 111-5) has been the most significant legislation that promotes and supports federal CCS RD&D program activities since passage of EISA. As discussed below, $3.4 billion in funding from the Recovery Act was intended to expand and accelerate the commercial deployment of CCS technologies to allow for commercial-scale demonstration in both new and retrofitted power plants and industrial facilities by 2020.

113th Congress

As introduced on January 9, 2014, H.R. 3826, the Electricity Security and Affordability Act, would essentially set requirements EPA must meet before the agency could issue greenhouse gas emission regulations under Section 111 of the Clean Air Act, such as the EPA re-proposed rule discussed above. On January 14, 2014, the Energy and Power Subcommittee, House Energy and Commerce Committee, voted to report the bill. Much of the discussion during the bill's markup centered on whether CCS was an adequately demonstrated technology to meet the requirements of the Clean Air Act. On January 28, 2014, the full committee voted to report the bill. Companion legislation, S. 1905, was introduced on January 9, 2014, and referred to the Senate Committee on Environment and Public Works.

A bill introduced on May 23, 2013, H.R. 2127, would prohibit the EPA from finalizing any rule limiting the emission of CO_2 from any existing or new source that is a fossil fuel-fired electric utility generating unit unless and until CCS becomes technologically and economically feasible. Other bills that target the EPA's proposed rule on limiting CO_2 emissions from new power plants were introduced shortly before and after the September 20, 2013, rule was proposed, including H.J.Res. 64, H.R. 3140, S. 1514, and others.

112th Congress

In the 112th Congress, a few bills were introduced that would have addressed aspects of CCS RD&D. The Department of Energy Carbon Capture and Sequestration Program Amendments Act of 2011 (S. 699) would have provided federal indemnification of up to $10 billion per project to early adopters of CCS technology (large CCS demonstration projects).[18] The New Manhattan Project for Energy Independence (H.R. 301) would have created a system of grants and prizes for a variety of technologies, including CCS, that would contribute to reducing U.S. dependence on foreign sources of energy. Other bills introduced would have provided tax incentives for the use of CO_2 in enhanced oil recovery (S. 1321), or would have eliminated the minimum capture requirement for the CO_2 sequestration tax credit (H.R. 1023). Other bills were also introduced that would have affected other aspects of CCS RD&D financing, such as loan guarantees. None of the bills introduced in the 112th Congress affecting federal CCS RD&D, other than the continuing resolution (CR), was enacted.

111th Congress

In the 111th Congress, two bills that would have authorized a national cap-and-trade system for limiting the emission of greenhouse gases (H.R. 2454 and S. 1733) also would have created programs aimed at accelerating the commercial availability of CCS. The programs would have generated funding from a surcharge on electricity delivered from the combustion of fossil fuels—approximately $1 billion per year—and made the funding available for grants, contracts, and financial assistance to eligible entities seeking to develop CCS technology.

Another source of funding in the bills was to come from a program that would distribute emission allowances to "early movers," entities that installed CCS technology on up to a total of 20 gigawatts of generating capacity. The House of Representatives passed H.R. 2454, but neither bill was enacted.

CCS RESEARCH, DEVELOPMENT, AND DEMONSTRATION: OVERALL GOALS

The U.S. Department of Energy states that the mission for the DOE Office of Fossil Energy is "to ensure the availability of ultra-clean (near-zero emissions), abundant, low-cost domestic energy from coal to fuel economic prosperity, strengthen energy security, and enhance environmental quality."[19] Over the past several years, the DOE Fossil Energy Research and Development Program has increasingly shifted activities performed under its Coal Program toward emphasizing CCS as the main focus.[20] The Coal Program represented 69% of total Fossil Energy Research and Development appropriations in FY2012 and in FY2013,[21] and represents nearly 70% in the FY2014 appropriation,[22] indicating that CCS has come to dominate coal R&D at DOE. This reflects DOE's view that "there is a growing consensus that steps must be taken to significantly reduce [greenhouse gas] emissions from energy use throughout the world at a pace consistent to stabilize atmospheric concentrations of CO_2, and that CCS is a promising option for addressing this challenge."[23]

DOE also acknowledges that the cost of deploying currently available CCS technologies is very high, and that to be effective as a technology for mitigating greenhouse gas emissions from power plants, the costs for CCS must be reduced. For example, in 2010 DOE stated that the cost of deploying available CCS post-combustion technology on a supercritical pulverized coal-fired power plant would increase the cost of electricity by 80%.[24] The challenge of reducing the costs of CCS technology is difficult to quantify, in part because there are no examples of currently operating commercial-scale coal-fired power plants equipped with CCS. Nor is it easy to predict when lower-cost CCS technology will be available for widespread deployment in the United States. Nevertheless, DOE observes that "the United States can no longer afford the luxury of conventional long-lead times for RD&D to bear results."[25] Thus the coal RD&D program is focused on achieving results that would allow for an advanced CCS technology portfolio to be ready by 2020 for large-scale demonstration.

The following section describes the components of the CCS activities within DOE's coal R&D program and their funding history since FY2012. This report focuses on this time period because during that time DOE obligated Recovery Act funding for its CCS programs, greatly expanding the CCS R&D portfolio. This was expected to accelerate the transition of CCS technology to industry for deployment and commercialization.[26] In addition, one remaining active project in the Clean Coal Power Initiative (CCPI) program that received funding in Round 2, prior to enactment of the Recovery Act—the Kemper County Energy Facility—is also discussed.

Program Areas

The 2010 RD&D *CCS Roadmap* described 10 different program areas pursued by DOE's Coal Program within the Office of Fossil Energy: (1) Innovations for Existing Plants (IEP); (2) Advanced Integrated Gasification Combined Cycle (IGCC); (3) Advanced Turbines; (4) Carbon Sequestration; (5) Solid State Energy Conversion Fuel Cells; (6) Fuels; (7) Advanced Research; (8) Clean Coal Power Initiative (CCPI); (9) FutureGen; and (10) Industrial Carbon Capture and Storage Projects (ICCS).[27]

DOE changed the program structure after FY2010, renaming and consolidating program areas. The program areas are divided into two main categories: (1) CCS Demonstration Programs, and (2) Carbon Capture and Storage, and Power Systems. **Table 1** shows the current program structure and indicates which programs received Recovery Act funding. In its FY2014 Budget Justification, DOE stated that the mission of these program areas is

> to support secure, affordable, and environmentally acceptable near-zero emissions fossil energy technologies. This will be accomplished via research, development, and demonstration to improve the performance of advanced CCS technologies. Commercial availability of CCS technologies will provide an option to use fossil fuel resources to provide provide energy and meet the President's climate goals.[28]

Some programs are directly focused on one or more of the three steps of CCS: capture, transportation, and storage. For example, the carbon capture program supports R&D on post-combustion, pre-combustion, and natural gas capture. The carbon storage program supports the regional carbon sequestration partnerships, geological storage technologies, and other aspects of permanently sequestering CO2 underground. In contrast, FutureGen from the outset was envisioned as combining all three steps: a zero-emission fossil fuel plant that would capture its emissions and sequester them in a geologic reservoir.

Table 1. DOE Carbon Capture and Storage Research, Development, and Demonstration Program Areas (funding in $ thousands, FY2012-FY2014, including Recovery Act funding)

Fossil Energy Research and Development Coal Program Areas	Program	Recovery Act	FY2012	FY2013 (annualized CR)[a]	FY2014 (Request)	P.L. 113-76
CCS Demonstrations	FutureGen 2.0	1,000,000	0	0	0	0
	Clean Coal Power Initiative (CCPI)	800,000	0	0	0	0
	Industrial Carbon Capture and Storage Projects (ICCS)	1,520,000	0	0	0	0
	Site Characterization, Training, Program Direction	80,000	0	0	0	0
Carbon Capture and Storage, and Power Systems	Carbon Capture	—	66,986	65,600	112,000	92,000
	Advanced Energy Systems	—	97,169	95,200	48,000	99,500
	Carbon Storage	—	112,208	109,900	61,095	108,900
	Cross Cutting Research	—	47,946	46,800	20,525	41,900
	NETL Coal Research and Development		35,011	33,300	35,011	50,000
		3,400,000	359,320	350,800	276,631	392,300

Source: U.S. Department of Energy, FY2013 Congressional Budget Request, volume 3, *Fossil Energy Research and Development*, http://energy.gov/sites/prod/files/2013/04 /f0/Volume3_1.pdf. U.S. Department of Energy, Carbon Sequestration, *Recovery Act*, http://www.fe.doe.gov/recovery/index.html; U.S. Department of Energy, FY2014. Congressional Budget Request, volume 3, *Fossil Energy Research and Development*, http://energy.gov/sites/prod/ files/2013/04/f0/Volume3_1.pdf; FY2014 Joint Explanatory Statement in the Congressional Record, January 15, 2014, and P.L. 113-76.

[a] According to DOE, the FY2013 column amounts reflect the continuing resolution (CR, P.L. 112-175) levels annualized to a full year. Figures reflect the March 1, 2013, sequester of funds under P.L. 112-25.

Within the CCS Demonstration Program Area, RD&D is also divided among different industrial sectors. The Clean Coal Power Initiative (CCPI) program area originally provided federal support to new coal technologies that helped power plants cut sulfur, nitrogen, and mercury pollutants. As CCS became the focus of coal RD&D, the CCPI program shifted to reducing greenhouse gas emissions by boosting plant efficiencies and capturing CO2.[29] In contrast, the ICCS program area demonstrates carbon capture technology for the non-power plant industrial sector.[30] Both these program areas focus on the *demonstration* component of RD&D, and account for $2.3 billion of the $3.4 billion appropriated for CCS RD&D in the Recovery Act in FY2009. From the budgetary perspective, the Recovery Act funding shifted the emphasis of CCS RD&D to large, industrial demonstration projects for carbon capture. The CCPI and ICCS program areas are discussed in more detail below.

This shift in emphasis to the demonstration phase of carbon capture technology is not surprising, and appears to heed recommendations from many experts who have called for large, industrial-scale carbon capture demonstration projects.[31] Primarily, the call for large-scale CCS demonstration projects that capture 1 million metric tons or more of CO2 per year reflects the need to reduce the additional costs to the power plant or industrial facility associated with capturing the CO2 before it is emitted to the atmosphere. The capture component of CCS is the costliest component, according to most experts.[32] The higher estimated costs to build and operate power plants with CCS compared to plants without CCS, and the uncertainty in cost estimates, results in part from a dearth of information about outstanding technical questions in carbon capture technology at the industrial scale.[33]

In comparative studies of cost estimates for other environmental technologies, such as for power plant scrubbers that remove sulfur and nitrogen compounds from power plant emissions (SO2 and NOx), some experts note that the farther away a technology is from commercial reality, the more uncertain is its estimated cost. At the beginning of the RD&D process, initial cost estimates could be low, but could typically increase through the demonstration phase before decreasing after successful deployment and commercialization. **Figure 1** shows a cost estimate curve of this type.

Deploying commercial-scale CCS demonstration projects—an emphasis within the DOE CCS RD&D program—would therefore provide cost estimates closer to operational conditions rather than laboratory- or pilot-plant-scale projects. In the case of SO_2 and NOx scrubbers, efforts typically took two decades or more to bring new concepts (such as combined SO_2 and NOx capture systems) to the commercial stage. As **Figure 1** indicates, costs for new technologies tend to fall over time with successful deployment and commercialization. It would be reasonable to expect a similar trend for CO_2 capture costs if the technologies become widely deployed.[34]

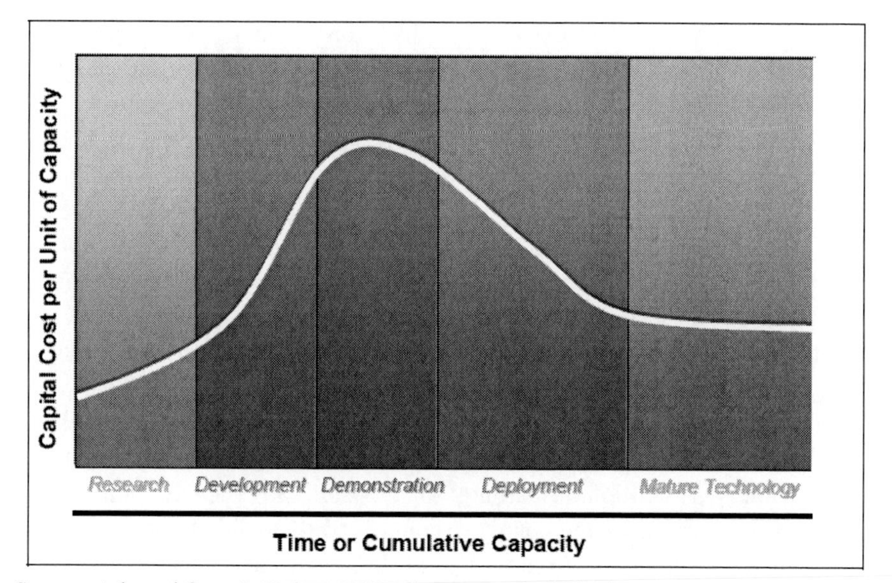

Source: Adapted from S. Dalton, "CO_2 Capture at Coal Fired Power Plants—Status and Outlook," 9th International Conference on Greenhouse Gas Control Technologies, Washington, DC, November, 16-20, 2008.

Figure 1. Typical Trend in Cost Estimates for a New Technology As It Develops from a Research Concept to Commercial Maturity.

First Full-Scale Project? The Kemper County Energy Facility

In a fact sheet accompanying the proposed rule limiting emissions of CO2 from new coal-fired power plants, the EPA asserts that CCS technology is currently feasible and refers to a coal-gasification project that is over 75% complete: the Kemper County Project. DOE awarded Southern Company Services a cooperative agreement under the CCPI Round 2 program, prior to enactment of the Recovery Act and the CCPI Round 3 awards, to develop technology at the Kemper County Energy Facility in Kemper County, Mississippi. The $270 million award was aimed to provide direct financial support for the development and deployment of a gasification technology called Transport Integrated Gasification (TRIG™).[35]

The Kemper County Project is an integrated gasification combined-cycle (IGCC)[36] power plant that will be owned and operated by Mississippi Power Company, a subsidiary of Southern Company, and which will use lignite as a fuel source. The plant is expected to have an estimated peak net output capability of 583 megawatts, and is designed to capture 65% of the total CO2 emissions released from the plant.[37]

According to DOE, this would make the CO2 emissions from the Kemper Project comparable to a natural gas-fired combined cycle power plant, and would therefore emit less than the 1,100 pounds per MWh limit as required by the new EPA proposed rule. The estimated 3 million tons of CO2 captured each year from the plant would be transported via newly constructed pipeline for use in enhanced oil recovery operations at nearby depleted oil fields in Mississippi. Commercial operation of the Kemper County Project is expected to begin in 2014, according to Southern Company.[38]

The $270 million award under Round 2 of the CCPI program represents less than 10% of the overall cost to build the plant, which is reported to be approximately $3 billion, according to a March 2013 DOE fact sheet.[39] However, in April 2013 the company announced that capital costs would be closer to $3.4 billion, approximately $1 billion higher than original cost estimates for the plant.[40]

In December 2013, Mississippi Power released documents indicating that the project was on schedule to begin operations in the last quarter of 2014, but that the total cost for the plant, including the lignite mine, CO2 pipeline, land purchase, and all the other components of the full project, was approximately $5 billion.[41]

It is likely that the plant will attract increased scrutiny in the wake of the EPA proposed rule on CO2 emissions, and its cost overruns evaluated against the promised environmental benefits due to CCS technology.[42] As **Figure 1** shows above, costs for technologies tend to peak for projects in the demonstration phase of development, such as the Kemper County Project. What the cost curve will look like, namely, how fast costs will decline and over what time period, is an open question and will likely depend on if and how quickly CCS technology is deployed on new and existing power plants.

RECOVERY ACT FUNDING FOR CCS PROJECTS: A LYNCHPIN FOR SUCCESS?

The bulk of Recovery Act funds for CCS ($3.32 billion, or 98%) was directed to three subprograms organized under the CCS Demonstrations Programs: the Clean Coal Power Initiative (CCPI), Industrial Carbon Capture and Storage projects (ICCS), and FutureGen (Table 1). Under the 2010 *CCS Roadmap*, and with the large infusion of funding from the Recovery Act, DOE's goal is to develop the technologies to allow for commercial-scale demonstration in both new and retrofitted power plants and industrial facilities by 2020. The DOE 2011 *Strategic Plan* sets a more specific target: to bring at least five commercial-scale CCS demonstration projects online by 2016.[43]

It could be argued that in its allocation of Recovery Act funding, DOE was heeding the recommendations of experts[44] who identified commercial-scale demonstration projects as the most important component, the lynchpin, for future development and deployment of CCS in the United States. It could also be argued that much of the future success of CCS is riding on these three programs. Accordingly, the following section provides a snapshot of the CCPI, ICCS, and FutureGen programs, and a brief discussion of some of their accomplishments and challenges.

CCS Demonstrations: CCPI, ICCS, and FutureGen 2.0

Clean Coal Power Initiative

The Clean Coal Power Initiative was an ongoing program prior to the $800 million funding increase from the Recovery Act. This funding now is being used to expand activities in this program area for CCPI Round 3 beyond developing technologies to reduce sulfur, nitrogen, and mercury pollutants from power plants.[45] After enactment of the Recovery Act, DOE did not request additional funding for CCPI under its Fossil Energy program in the annual appropriations process (**Table 1** shows zero dollars for FY2012-FY2014). Rather, in the FY2010 DOE budget justification, DOE stated that funding for the these projects in CCPI Round 3 would be supported through the Recovery Act, and as a result "DOE will make dramatic progress in demonstrating CCS at commercial scale using these funds without the need for additional resources for demonstration in 2010."[46]

Table 2. DOE CCS Demonstration Round 3 Projects

Round 3 Project	Location	DOE Share of Funding ($ millions)	Total Project Cost ($ millions)	Percent DOE Share	Metric Tons of CO_2 Captured Annually (millions)	Project Status
Texas Clean	Penwell, TX	450	1,727	26%	2.7b	Active
Energy Project						
Hydrogen Energy California Project	Kern County, CA	408	4,028	10%	2.6	Active
NRG Energy Project	Thompsons, TX	167	338	50%	1.4	Active
AEP Mountaineer Project	New Haven,WV	334	668	50%	1.5	Withdrawn
Southern	Mobile, AL	295	665	44%	1	Withdrawn
Company Project						
Bas-in Electric	Beulah, ND	100	387	26%	0.9	Withdrawn
Power Project						
Total		1,754	7,813	22.4%	10.1	
Total, Active		1,025	6,093	16.8%	6.7	
Projects[a]						

Sources: DOE Fossil Energy Techline; Environment News Service (March 12, 2010), http://www.ensnewswire.com/ens/mar2010/2010-03-12-093.html; NETL CCPI website, http://www.netl.doe.gov/technologies/ coalpower/cctc/ccpi/index.html; NETL Factsheet: Summit Texas Clean Energy, LLC, March 2012, http:// www.netl.doe.gov/publications/factsheets/project/FE0002650.pdf; NETL Factsheet Hydrogen Energy California Project, May 2013, http://www.netl. doe.gov/publications/factsheets/project/FE0000663.pdf; NETL Factsheet BRG Energy: W.A. Parish Post Combustion CO2 Capture and Sequestration Project, March 2012, http://www.netl.doe.gov/publications /factsheets/project/FE0003311. pdf.

Notes: DOE funding for the NRG Energy Project was initially announced as up to $154 million (see March 9, 2009, DOE Techline, http://www.fossil.energy.gov /news/techlines/2010/10005- NRG_Energy_Selected_to_Receive_DOE.html). A May 2010 DOE fact sheet indicates that funding for NRG is $167 million (http://www.netl.doe.gov /publications/factsheets/project/FE0003311.pdf).

[a] Total include amounts that were reallocated from withdrawn projects to active projects.

[b] According to NETL, this amount could be up to 3 million metric tons annually.

According to the 2010 DOE *CCS Roadmap*, Recovery Act funds are being used for these demonstration projects to "allow researchers broader CCS commercial-scale experience by expanding the range of technologies, applications, fuels, and geologic formations that are being tested."[47] DOE selected six projects under CCPI Round 3 through two separate solicitations.[48]

> The total DOE share of funding would have been $1.75 billion for the six projects in five states: Alabama, California, North Dakota, Texas, and West Virginia (**Table 2**). However, the projects in Alabama, North Dakota, and West Virginia withdrew from the program, and currently the DOE share for the remaining three projects is $1.03 billion (of a total of over $6 billion for total expected costs). With the withdrawal of three CCPI Round 3 projects, DOE's share of the total program costs shrank from over 22% to approximately 17%.

Reasons for Withdrawal from the CCPI Program

Commercial sector partners identified a number of reasons for withdrawing from the CCPI program, including finances, uncertainty regarding future regulations, and uncertainty regarding the future national climate policy.

Southern Company—Plant Barry 160 MW Project: Southern Company withdrew its Alabama Plant Barry project from the CCPI program on February 22, 2010, slightly more than two months after DOE Secretary Chu announced $295 million in DOE funding for the 11-year, $665 million project that would have captured up to 1 million tons of CO_2 per year from a 160 MW coal-fired generation unit.[49] According to some sources, Southern Company's decision was based on concern about the size of the company's needed commitment (approximately $350 million) to the project, and its need for more time to perform due diligence on its financial commitment, among other reasons.[50] Southern Company continues work on a much smaller CCS project that would capture CO_2 from a 25 MW unit at Plant Barry.

Basin Electric Power—Antelope Valley 120 MW Project: On July 1, 2009, Secretary Chu announced $100 million in DOE funding for a project that would capture approximately 1 million tons of CO_2 per year from a 120 MW electric-equivalent gas stream from the Antelope Valley power station near Beulah, ND.[51] In December 2010, the Basin Electric Power Cooperative withdrew its project from the CCPI program, citing regulatory uncertainty with regard to capturing CO_2, uncertainty about the project's cost (one source indicates that the company estimated $500 million total cost; DOE estimated $387 million—see **Table 2**),[52] uncertainty of environmental legislation, and lack of a long-term energy strategy for the country.[53] The project would have supplied the captured CO_2 to an existing pipeline that transports CO_2 from the Great Plains Synfuels Plant near Beulah for enhanced oil recovery in Canada's Weyburn field approximately 200 miles north in Saskatchewan.

American Electric Power—Mountaineer 235 MW Project: In July 2011 American Electric Power decided to halt its plans to build a carbon capture plant for a 235 MW generation unit at its 1.3 gigawatt Mountaineer power plant in New Haven, WV. The project represented Phase 2 of an ongoing CCPI project. Secretary Chu had earlier announced a $334 million award for the project on December 4, 2009.[54] According to some sources, AEP dropped the project because the company was not certain that state regulators would allow it to recover the additional costs for the CCS project through rate increases charged to its customers.[55] In addition, company officials cited broader economic and policy conditions as reasons for cancelling the project.[56] Some commentators suggested that congressional inaction on setting limits on greenhouse gas emissions, as well as the weak economy, may have diminished the incentives for a company like AEP to invest in CCS.[57] One source concluded that "Phase 2 has been cancelled due to unknown climate policy."[58]

Reshuffling of Funding for CCPI

According to DOE, $140 million of the $295 million previously allotted to the Southern Company Plant Barry project was redistributed to the Texas Clean Energy project and the Hydrogen Energy California project. DOE provided additional funding, resulting in each project receiving an additional $100 million above its initial awards.[59] The remaining funding from the canceled Plant Barry project (up to $154 million) was allotted to the NRG Energy project in Texas (see **Table 2**).[60]

According to a DOE source, selection of the Basin Electric Power project was announced but a cooperative agreement was never awarded by DOE.[61] Funds that were to be obligated for the Basin project could therefore have been reallocated within the department, but were rescinded by Congress in FY2011 appropriations.

Some of the funding for the AEP Mountaineer project was rescinded by Congress in FY2012 appropriations legislation (P.L. 112-74). In the report accompanying P.L. 112-74, Congress rescinded a total of $187 million of prior-year balances from the Fossil Energy Research and Development account.[62] The rescission did not apply to amounts previously appropriated under P.L. 111-5; however, funding for the AEP Mountaineer project that was provided by the Recovery Act and not spent was returned to the Treasury and not made available to the CCPI program.[63]

Industrial Carbon Capture and Storage Projects

The original DOE ICCS program was divided into two main areas: Area 1, consisting of large industrial demonstration projects; and Area 2, consisting of projects to test innovative concepts for the beneficial reuse of CO2.[64] Under Area 1, the first phase of the program consisted of 12 projects cost-shared with private industry, intended to increase investment in clean industrial technologies and sequestration projects. Phase 1 projects averaged approximately seven months in duration. Following Phase 1, DOE selected three projects for Phase 2 for design, construction, and operation.[65] The three Phase 2 projects are listed as large-scale demonstration projects in **Table 3**. The total share of DOE funding for the three projects, provided by the Recovery Act, is $686 million, or approximately 64% of the sum total Area 1 program cost of $1.075 billion.

Under Area 2, the initial phase consisted of $17.4 million in Recovery Act funding and $7.7 million in private-sector funding for 12 projects to engage in feasibility studies to examine the beneficial reuse of CO2.[66] In July 2010, DOE selected six projects from the original 12 projects for a second phase of funding to find ways of converting captured CO2 into useful products such as fuel, plastics, cement, and fertilizer. The six projects are listed under "Innovative Concepts/Beneficial Use" in **Table 3**. The total share of DOE funding for the six projects, provided by the Recovery Act, is $141.5 million, or approximately 71% of the sum total cost of $198.2 million.

Since its original conception, the DOE ICCS program has expanded with an additional 22 projects, funded under the Recovery Act, to accelerate promising technologies for CCS.[67] In its listing of the 22 projects, DOE groups them into four general categories: (1) Large-Scale Testing of Advanced Gasification Technologies; (2) Advanced Turbo-Machinery to Lower Emissions from Industrial Sources; (3) Post-Combustion CO2 Capture with Increased Efficiencies and Decreased Costs; and (4) Geologic Storage Site Characterization.[68] The total share of DOE funding for the 22 projects, provided by the Recovery Act, is $594.9 million, or approximately 78% of the sum total cost of $765.2 million.

Overall, the total share of federal funding for all the ICCS projects combined is $1.422 billion, or approximately 70% of the sum total cost of $2.038 billion.

Table 3. DOE Industrial Carbon Capture and Storage (ICCS) Projects (showing DOE share of funding and total project cost)

ICCS Project Name	Location	Type of Project	DOE Share of Funding ($ millions)	Total Project Cost ($ millions)	Percent DOE Share
Air Products & Chemicals, Inc.	Port Arthur, TX	Large-Scale Demonstration	284	431	66%
Archer Daniels Midland Co.	Decatur, IL	Large-Scale Demonstration	141	208	68%
Leucadia Energy, LLC	Lake Charles, LA	Large-Scale Demonstration	261	436	60%
Alcoa, Inc.	Alcoa Center, PA	Innovative Concepts/Beneficial Use	13.5	16.9	80%
Novomer, Inc.	Ithaca, NY	Innovative Concepts/Beneficial Use	20.5	25.6	80%
Touchstone Research Lab, Ltd.	Triadelphia, PA	Innovative Concepts/Beneficial Use	6.7	8.4	80%
Phycal, LLC	Highland Heights, OH	Innovative Concepts/Beneficial Use	51.4	65	80%
Skyonic Corp.	Austin, TX	Innovative Concepts/Beneficial Use	28	39.6	70%
Calera Corp.	Los Gatos, CA	Innovative Concepts/Beneficial Use	21.4	42.7	50%
Air Products & Chemicals, Inc.	Allentown, PA	Advanced Gasification Technologies	71.7	75	96%
Eltron Research & Development, Inc.	Boulder, CO	Advanced Gasification Technologies	71.4	73.7	97%

ICCS Project Name	Location	Type of Project	DOE Share of Funding ($ millions)	Total Project Cost ($ millions)	Percent DOE Share
Research Triangle Institute	Research Triangle Park, NC	Advanced Gasification Technologies	168.8	174	97%
GE Energy	Schenectady, NY	Advanced Turbo-Machinery	31.3	62.6	50%
Siemens Energy	Orlando, FL	Advanced Turbo-Machinery	32.3	64.7	50%
Clean Energy Systems, Inc.	Rancho Cordova, CA	Advanced Turbo-Machinery	30	42.9	70%
Ramgen Power Systems	Bellevue, WA	Advanced Turbo-Machinery	50	79.7	63%
ADA-ES, Inc.	Littleton, CO	Post-Combustion Capture	15	18.8	80%
Alstom Power	Windsor, CT	Post-Combustion Capture	10	12.5	80%
Membrane Technology & Research, Inc.	Menlo Park, CA	Post-Combustion Capture	15	18.8	80%
Praxair	Tonawanda, NY	Post-Combustion Capture	35	55.6	63%
Siemens Energy, Inc.	Pittsburgh, PA	Post-Combustion Capture	15	18.8	80%
Board of Trustees U. of IL	Champaign, IL	Geologic Site Characterization	5	6.5	77%
N. American Power Group, Ltd.	Greenwood Village, CO	Geologic Site Characterization	5	7.85	64%
Sandia Technologies, LLC	Houston, TX	Geologic Site Characterization	4.38	5.63	78%
S. Carolina Research Foundation	Columbia, SC	Geologic Site Characterization	5	6.25	80%
Terralog Technologies USA, Inc.	Arcadia, CA	Geologic Site Characterization	5	6.25	80%
U. of Alabama	Tuscaloosa, AL	Geologic Site Characterization	5	10.8	46%

Table 3. (Continued)

ICCS Project Name	Location	Type of Project	DOE Share of Funding ($ millions)	Total Project Cost ($ millions)	Percent DOE Share
U. of Kansas Center for Research, Inc.	Lawrence, KS	Geologic Site Characterization	5	6.29	80%
U. of Texas at Austin	Austin, TX	Geologic Site Characterization	5	6.25	80%
U. of Utah	Salt Lake City, UT	Geologic Site Characterization	5	7.23	69%
U. of Wyoming	Laramie, WY	Geologic Site Characterization	5	5	100%
		Totals	1,422.4	2,038.4	70%

Source: Emails from Regis K. Conrad, Director, Division of Cross-Cutting Research, DOE, March 20 and March 27, 2012; U.S. DOE, National Energy Technology Laboratory, *Major Demonstrations, Industrial Capture and Storage (ICCS): Area 1*, http://www.netl.doe.gov/technologies/coalpower/cctc/iccs1/index.html#; U.S. DOE, *Carbon Capture and Storage from Industrial Sources, Industrial Carbon Capture Project Selections,* http://fossil.energy.gov/recovery/projects /iccs_projects_0907101.pdf.

Notes: Table is ordered from top to bottom by type of project: Large-Scale Demonstration; Innovative Concepts/Beneficial Use; Advanced Gasification Technologies; Advanced Turbo-Machinery; Post-Combustion Capture; and Geologic Site Characterization. Totals may not add due to rounding.

FutureGen—A Special Case?

On February 27, 2003, President George W. Bush proposed a 10-year, $1 billion project to build a coal-fired power plant that would integrate carbon sequestration and hydrogen production at a 275 megawatt-capacity plant, enough to power about 150,000 average U.S. homes. As originally conceived, the plant would have been a coal-gasification facility and would have produced and sequestered between 1 million and 2 million tons of CO2 annually. On January 30, 2008, DOE announced that it was "restructuring" the FutureGen program away from a single, state-of-the-art "living laboratory" of integrated R&D technologies—a single plant—to instead pursue a new strategy of multiple commercial demonstration projects.[69] In the restructured program, DOE would support up to two or three demonstration projects of at least 300 megawatts that would sequester at least 1 million tons of CO2 per year.

In the Bush Administration's FY2009 budget, DOE requested $156 million for the restructured FutureGen program, and specified that the federal cost-share would only cover the CCS portions of the demonstration projects, not the entire power system. However, after the Recovery Act was enacted on February 17, 2009, Secretary Chu announced an agreement with the FutureGen Alliance—an industry consortium—to advance construction of the FutureGen plant built in Mattoon, IL, the site selected by the FutureGen Alliance in 2007.[70] Further, DOE anticipated that $1 billion of funding from the Recovery Act would be used to support the project.[71]

On August 5, 2010, then-Secretary of Energy Chu announced the $1 billion award, from Recovery Act funds, to the FutureGen Alliance, Ameren Energy Resources, Babcock & Wilcox, and Air Liquide Process & Construction, Inc., to build FutureGen 2.0.[72] FutureGen 2.0 differs from the original concept for the plant, because it would retrofit Ameren's existing power plant in Meridosia, IL, with oxy-combustion technology at a 202 MW oil-fired unit,[73] rather than build a new state-of-the-art plant in Mattoon.[74]

Challenges to FutureGen—A Similar Path for Other Demonstration Projects?

More than a decade after the George W. Bush Administration announced FutureGen—its signature clean coal power initiative—the program is still in early development. Among the challenges to the development of FutureGen 2.0 are rising costs of production, ongoing issues with project development, lack of incentives for investment from the private sector, time constraints, and competition with foreign nations. Remaining challenges to FutureGen's development include securing private sector funding to meet increasing costs, purchasing the power plant for the project, obtaining permission from DOE to retrofit the plant, performing the retrofit, and then meeting the goal of 90% capture of CO_2.[75]

A question for Congress is whether FutureGen represents a unique case of a first mover in a complex, expensive, and technically challenging endeavor, or whether it represents all large CCS demonstration projects once they move past the planning stage. As discussed above, approximately $3.3 billion of Recovery Act funding is committed to large demonstration projects, including FutureGen. A rationale for committing such a substantial level of funding to demonstration projects was to scale up CCS RD&D more quickly than had been the pace of technology development prior to enactment of the Recovery Act. However, if all the CCS demonstration projects encounter similar changes in scope, design, location, and cost as FutureGen, the goals laid out in the DOE 2011 *Strategic Plan*—namely, to bring at least five commercial-scale CCS demonstration projects online by 2016—may be in jeopardy.

Alternatively, one could argue that FutureGen from its original conception was unique. None of the other large-scale demonstration projects share the same original ambitious vision: to create a new, one-of-a-kind, near-zero emission CCS plant from the ground up. Even though the individual components of FutureGen—as originally conceived—were not themselves new innovations, combining the capture, transportation, and storage components together into a 250- megawatt functioning power plant could be considered unprecedented and therefore most likely to experience delays at each step in development.

Scholars have described the stages of technological change in different schemes, such as

- invention, innovation, adoption, diffusion;[76] or
- technology readiness levels (TRLs) ranging from TRL 1 (basic technology research) to TRL 9 (system test, launch, and operations);[77] or
- conceptual design, laboratory/bench scale, pilot plant scale, full-scale demonstration plant, and commercial process.[78]

FutureGen is difficult to categorize within these schemes, in part because the project spanned a range of technology development levels irrespective of the particular scheme. The original conception of the FutureGen project arguably had aspects of conceptual design through commercial processes—all five components of the scheme listed as the third bullet above—which meant that the project was intended to march through all stages in a linear fashion. As some scholars have noted, however, the stages of technological change are highly interactive, requiring learning by doing and learning by using, once the project progresses past its innovative stage into larger-scale demonstration and deployment.[79] The task of tackling all the stages of technology development in one project—the original FutureGen—might have been too daunting and, in addition to other factors, contributed to the project's erratic progress since 2003. It remains to be seen whether the current large-scale demonstration projects funded by DOE under CCPI Round 3 follow the path of FutureGen or instead achieve their technological development goals on time and within their current budgets. Presumably, lessons learned during the planning, construction, and operation of these demonstration projects will be shared with the broader electric power industry.[80]

GEOLOGIC SEQUESTRATION/STORAGE: DOE RD&D FOR THE LAST STEP IN CCS

DOE has allocated $112 million in FY2012, $110 million in FY2013, and $109 million in FY2014 for its carbon sequestration/storage activities. (See Table 1.) In contrast with the carbon capture technology RD&D, which received nearly all of the $3.4 billion from Recovery Act funding, carbon sequestration/carbon storage activities received approximately $50 million in Recovery Act funds. Recovery Act funds were awarded for 10 projects to conduct site characterization of promising geologic formations for CO2 storage.[81]

Brief History of DOE Geological Sequestration/Storage Activities

DOE has devoted the bulk of its funding for geological sequestration/ storage activities to RD&D efforts for injecting CO2 into subsurface geological reservoirs. Injection and storage is the third step in the CCS process following the CO2 capture step and CO2 transport step. One part of the RD&D effort is characterizing geologic reservoirs (which received a $50 million boost from Recovery Act funds, as noted above); however, the overall program is much broader than just characterization, and has now reached the beginning of the phase of large-volume CO2 injection demonstration projects across the country. According to DOE, these large-volume tests are needed to validate long-term storage in a variety of different storage formations of different depositional environments, including deep saline reservoirs, depleted oil and gas reservoirs, low permeability reservoirs, coal seams, shale, and basalt.[82] The large-volume tests can be considered injection experiments conducted at a commercial scale (i.e., approximately 1 million tons of CO2 injected per year) that should provide crucial information on the suitability of different geologic reservoirs; monitoring, verification, and accounting of injected CO2; risk assessment protocols for long-term injection and storage; and other critical challenges.

In 2003 DOE created seven regional carbon sequestration partnerships (RCSPs), essentially consortia of public and private sector organizations grouped by geographic region across the United States and parts of Canada.[83] The geographic representation was intended to match regional differences in fossil fuel use and geologic reservoir potential for CO2 storage.[84] The RCSPs cover 43 states and 4 Canadian provinces and include over 400 organizations, according to the DOE 2011 *Strategic Plan*. **Table 4** shows the seven partnerships, the lead organization for each, and the states and provinces included. Several states belong to more than one RCSP.

The RCSPs have pursued their objectives through three phases beginning in 2003:

(1) Characterization Phase (2003 to 2005), an initial examination of the region's potential for geological sequestration of CO2; (2) Validation Phase (2005 to 2011), small-scale injection field tests (less than 500,000 tons of CO2) to develop a better understanding of how different geologic formations would handle large amounts of injected CO2; and (3) Development Phase (2008 to 2018 and beyond), injection tests of at least 1 million tons of CO2 to simulate commercial-scale quantities of injected CO2.[85] The last phase is intended also to collect enough information to help understand the regulatory, economic, liability, ownership, and public outreach requirements for commercial deployment of CCS.

There are RD&D activities funded by DOE under its carbon sequestration/carbon storage program activities other than the RCSPs, such as geological storage technologies; monitoring, verification, and assessment; carbon use and reuse; and others. However, the RCSPs were allocated approximately 70% of annual spending on carbon sequestration/carbon storage in FY2012, and comprised 66% of that account in the FY2014 budget request.

The RCSPs provide the framework and infrastructure for a wide variety of DOE geologic sequestration/storage activities.

Table 4. Regional Carbon Sequestration Partnerships

Regional Carbon Sequestration Partnership (RCSP)	Lead Organization	States and Provinces in the Partnership
Big Sky Carbon Sequestration Partnership (BSCSP)	Montana State University-Bozeman	MT, WY, ID, SD, eastern WA, eastern OR
Midwest Geological Sequestration Consortium (MGSC)	Illinois State Geological Survey	IL, IN, KY
Midwest Regional Carbon Sequestration Partnership (MRCSP)	Battelle Memorial Institute	IN, KY, MD, MI, NJ, NY, OH, PA, WV,
Plains CO2 Reduction Partnership (PCOR)	University of North Dakota Energy and Environmental Research Center	MT, northeast WY, ND, SD, NE, MN, IA, MO, WI, Manitoba, Alberta, Saskatchewan, British Columbia (Canada)
Southeast Regional Carbon Sequestration Partnership (SECARB)	Southern States Energy Board	AL, AS, FL, GA, LA, MS, NC, SC, TN, TX, VA, portions of KY and WV
Southwest Regional Partnership on Carbon Sequestration (SWP)	New Mexico Institute of Mining and Technology	AZ, CO, OK, NM, UT, KS, NV, TX, WY
West Coast Regional Carbon Sequestration Partnership (WESTCARB)	California Energy Commission	AK, AZ, CA, HI, OR, NV, WA, British Columbia (Canada)

Source: DOE National Energy Technology Laboratory, *Carbon Sequestration Regional Carbon Sequestration Partnerships*, http://www.netl.doe.gov /technologies/carbon_seq/infrastructure/rcsp.html.

Current Status and Challenges to Carbon Sequestration/Storage

The third phase—Development—is currently underway for all the RCSPs, and large-scale CO2 injection has begun for the SECARB and MGSC projects.[86] The Development Phase large-scale injection projects are arguably akin to the large-scale carbon capture demonstration projects discussed above

(See **Table 2**). They are needed to understand what actually happens to CO2 underground when commercial-scale volumes are injected in the same or similar geologic reservoirs as would be used if CCS were deployed nationally.

In addition to understanding the technical challenges to storing CO2 underground without leakage over hundreds of years, DOE also expects that the Development Phase projects will provide a better understanding of regulatory, liability, and ownership issues associated with commercial-scale CCS.[87] These nontechnical issues are not trivial, and could pose serious challenges to widespread deployment of CCS even if the technical challenges of injecting CO2 safely and in perpetuity are resolved. For example, a complete regulatory framework for managing the underground injection of CO2 has not been developed in the United States. However, EPA promulgated a rule under the authority of the Safe Drinking Water Act (SDWA) that creates a new class of injection wells under the existing Underground Injection Control Program. The new class of wells (Class VI) establishes national requirements specifically for injecting CO2 and protecting underground sources of drinking water. EPA's stated purpose in proposing the rule was to ensure that CCS can occur in a safe and effective manner in order to enable commercial-scale CCS to move forward.[88]

The development of the regulation for Class VI wells highlighted that EPA's authority under the SDWA is limited to protecting underground sources of drinking water but does not address other major issues. Some of these include the long-term liability for injected CO2, regulation of potential emissions to the atmosphere, legal issues if the CO2 plume migrates underground across state boundaries, private property rights of owners of the surface lands above the injected CO2 plume, and ownership of the subsurface reservoirs (also referred to as pore space).[89] Because of these issues and others, there are some indications that broad community acceptance of CCS may be a challenge.

The large-scale injection tests may help identify the key factors that lead to community concerns over CCS, and help guide DOE, EPA, other agencies, and the private sector towards strategies leading to the widespread deployment of CCS. Currently, however, the general public is largely unfamiliar with the details of CCS and these challenges have yet to be resolved.[90]

OUTLOOK

Testimony from Scott Klara of the National Energy Technology Laboratory sums up a crucial metric for the success of the federal CCS RD&D program, namely, whether CCS technologies are deployed in the commercial marketplace:

> The success of the Clean Coal Program will ultimately be judged by the extent to which emerging technologies get deployed in domestic and international marketplaces. Both technical and financial challenges associated with the deployment of new "high risk" coal technologies must be overcome in order to be capable of achieving success in the marketplace. Commercial scale demonstrations help the industry understand and overcome startup issues, address component integration issues, and gain the early learning commercial experience necessary to reduce risk and secure private financing and investment for future plants.[91]

To date, there are no commercial ventures in the United States that capture, transport, and inject large quantities of CO_2 (e.g., 1 million tons per year or more) solely for the purposes of carbon sequestration.

However, the CCS RD&D program has embarked on commercial-scale demonstration projects for CO_2 capture, injection, and storage. The success of these demonstration projects will likely bear heavily on the future outlook for widespread deployment of CCS technologies as a strategy for preventing large quantities of CO_2 from reaching the atmosphere while plants continue to burn fossil fuels, mainly coal. The September 20, 2013, re-proposal of an EPA standard to limit CO_2 emissions from coal-fired power plants has invited renewed scrutiny of CCS technology and its prospects for commercial deployment. Congress may wish to carefully review the results from these demonstration projects as they progress in order to gauge whether DOE is on track to meet its goal of allowing for an advanced CCS technology portfolio to be ready by 2020 for large-scale demonstration and deployment in the United States.

In addition to the issues and programs discussed above, other factors might affect the demonstration and deployment of CCS in the United States. The use of hydraulic fracturing techniques to extract unconventional natural gas deposits recently has drawn national attention to the possible negative consequences of deep well injection of large volumes of fluids. Hydraulic fracturing involves the high-pressure injection of fluids into the target formation to fracture the rock and release natural gas or oil. The injected fluids, together with naturally occurring fluids in the shale, are referred to as produced water. Produced waters are pumped out of the well and disposed of. Often the produced waters are disposed of by re-injecting them at a different site in a different well. These practices have raised concerns about possible leakage as fluids are pumped into and out of the ground, and about deep-well injection causing earthquakes. Public concerns over hydraulic fracturing and deep-well injection of produced waters may spill over into concerns about deep-well injection of CO_2. How successfully DOE is able to address these types of concerns as the large-scale demonstration projects move forward into their injection phases could affect the future of CCS deployment.

End Notes

[1] U.S. Department of Energy, National Energy Technology Laboratory, *Carbon Sequestration Program: Technology Program Plan*, Enhancing the Success of Carbon Capture and Storage Technologies, February 2011, p. 10, http://www.netl.doe.gov/technologies /carbon_seq/refshelf/2011_Sequestration_Program_Plan.pdf.

[2] U.S. Department of Energy, National Energy Technology Laboratory, *DOE/NETL Carbon Dioxide Capture and Storage RD&D Roadmap*, December 2010. Hereinafter referred to as the DOE 2010 *CCS Roadmap*. See http://www.netl.doe.gov/File%20Library/Research /Carbon%20Seq/Reference%20Shelf/CCSRoadmap.pdf.

[3] As originally conceived in 2003, FutureGen would have been a 10-year project to build a coal-fired power plant that would integrate carbon sequestration and hydrogen production while producing 275 megawatts of electricity, enough to power about 150,000 average U.S. homes. The plant would have been a coal-gasification facility and would have produced and sequestered between 1 million and 2 million tons of CO2 annually. FutureGen 2.0 differs from the original concept for the plant, because it would retrofit an existing power plant in Meredosia, IL, with oxy-combustion technology, and is funded largely by appropriations made available by the Recovery Act. See CRS Report R43028, *The FutureGen Carbon Capture and Sequestration Project: A Brief History and Issues for Congress*, by Peter Folger.

[4] See, for example, CRS Report R42532, *Carbon Capture and Sequestration (CCS): A Primer;* CRS Report R41325, *Carbon Capture: A Technology Assessment.*

[5] For a fuller discussion of the proposed rule and EPA standards for greenhouse gas emissions from power plants, see CRS Report R43127, *EPA Standards for Greenhouse Gas Emissions from Power Plants: Many Questions, Some Answers*, by James E. McCarthy.

[6] The proposal and background information is available at http://www2.epa.gov/carbon-pollution-standards/2013- proposed-carbon-pollution-standard-new-power-plants.

[7] Environmental Protection Agency, "Standards of Performance for Greenhouse Gas Emissions From New Stationary Sources: Electric Utility Generating Units," 79 *Federal Register* 1429, January 8, 2014.

[8] Ibid.

[9] DOE 2010 *CCS Roadmap*, p. 3.

[10] Office of the Press Secretary, The White House, "Power Sector Carbon Pollution Standards," Memorandum for the Administrator of the Environmental Protection Agency, June 25, 2013, http://www.whitehouse.gov/the-press-office/ 2013/06/25/presidential-memorandum-power-sector-carbon-pollution-standards.

[11] For a detailed discussion of the EPA's regulation of coal, see CRS Report R41914, *EPA's Regulation of Coal-Fired Power: Is a "Train Wreck" Coming?*, by James E. McCarthy and Claudia Copeland.

[12] The projects cited in the re-proposed rule are the Southern Company Kemper County Energy Facility, the SaskPower Boundary Dam CCS project, the Summit Power Texas Clean Energy Project, and the Hydrogen Energy California Project. The Boundary Dam project is a Canadian venture; the other three projects are in the United States and are receiving funding from DOE.

[13] For a detailed discussion of how natural gas is affecting electric power generation, see CRS Report R42814, *Natural Gas in the U.S. Economy: Opportunities for Growth* , by Robert Pirog and Michael Ratner.

[14] The other three are feasibility, costs, and size of emission reductions.

[15] See for example, the November 15, 2013, letter to EPA Administrator Gina McCarthy from Representative Fred Upton, chair of the House Committee on Energy and Commerce, http://www.eenews.net/assets/2013/11/22/ document_daily_03.pdf; and the December 19, 2013, letter to Administrator McCarthy from Representative Lamar Smith, chair of the House Committee on Science, Space, and Technology, http://science.house.gov/sites/ republicans.science.house.gov/files/documents/Letters/121913_mccarthy.pdf.

[16] P.L. 109-58, Title IX, Subtitle F, §963; 42 U.S.C. 16293.

[17] P.L. 110-140, Title VII, Subtitles A and B.

[18] Among other provisions, the bill would also have amended EISA to expand the number of large CCS demonstration projects from 7 to 10.

[19] DOE 2010 *CCS Roadmap*, p. 2.

[20] The Coal Program contains CCS RD&D activities, and is within DOE's Office of Fossil Energy, Fossil Energy Research and Development, as listed in DOE detailed budget justifications for each fiscal year. See, for example, U.S. Department of Energy, FY2014 Congressional Budget Request, volume 3, *Fossil Energy Research and Development*, http://energy.gov/sites/prod/files/2013/04/f0/Volume3_1.pdf. The percentage of funding allocated to the Coal Program is calculated based on the subtotal for Fossil Energy Research and Development prior to rescission of prior year balances, which were $187 million for FY2012 and $42 million for FY2013, respectively.

[21] U.S. Department of Energy, FY2013 Congressional Budget Request, volume 3, *Fossil Energy Research and Development*, p. 411.

[22] FY2014 Joint Explanatory Statement in the Congressional Record, January 15, 2014, and P.L. 113-76.

[23] DOE 2010 *CCS Roadmap*, p. 3.

[24] DOE 2010 *CCS Roadmap*, p. 3.

[25] DOE 2010 *CCS Roadmap*, p. 3.

[26] DOE 2010 *CCS Roadmap*, p. 2.

[27] DOE 2010 *CCS Roadmap*, p. 11.

[28] U.S. Department of Energy, FY2014 Congressional Budget Request, volume 3, *Fossil Energy Research and Development*, p. FE-13.

[29] Ibid., p. FE-16.

[30] DOE 2010 *CCS Roadmap*, p. 12.

[31] See, for example, the presentations given by Edward Rubin of Carnegie Mellon University, Howard Herzog of the Massachusetts Institute of Technology, and Jeff Phillips of the Electric Power Research Institute, at the CRS seminar *Capturing Carbon for Climate Control: What's in the Toolbox and What's Missing*, November 18, 2009. (Presentations available from Peter Folger at 7-1517.) Rubin stated that at least 10 full-scale demonstration projects would be needed to establish thc reliability and true cost of CCS in power plant applications. Herzog also called for at least 10 demonstration plants worldwide that capture and sequester a million metric tons of CO2 per year. In his presentation, Phillips stated that large-scale demonstrations are critical to building confidence among power plant owners.

[32] For example, an MIT report estimated that the costs of capture could be 80% or more of the total CCS costs. John Deutsch et al., *The Future of Coal*, Massachusetts Institute of Technology, An Interdisciplinary MIT Study, 2007, Executive Summary, p. xi.

[33] *The Future of Coal*, p. 97.

[34] For a fuller discussion of the relationship between costs of developing technologies analogous to CCS, such as SO2 and NOx scrubbers, see CRS Report R41325, *Carbon Capture: A Technology Assessment*, by Peter Folger.

[35] NETL Fact Sheet, "Demonstration of a Coal-Based Transport Gasifier," Fact Sheet NT42391, March 2013, http://www.netl.doe.gov/publications/factsheets/project/NT42391.pdf.

[36] For more information on IGCC power plants and CCS, see CRS Report R41325, *Carbon Capture: A Technology Assessment*, by Peter Folger.

[37] NETL Fact Sheet, "Demonstration of a Coal-Based Transport Gasifier," Fact Sheet NT42391, March 2013, http://www.netl.doe.gov/publications/factsheets/project/NT42391.pdf.

[38] Mississippi Power, A Southern Company, "Kemper County Energy Facility, Kemper Timeline," http://www.mississippipower.com/kemper/project-timeline.asp.

[39] NETL Fact Sheet, "Demonstration of a Coal-Based Transport Gasifier," Fact Sheet NT42391, March 2013, http://www.netl.doe.gov/publications/factsheets/project/NT42391.pdf.

[40] Tamar Hallerman, "Miss. Power to Absorb $540M in Cost Increases from Kemper Plant," *GHG Reduction Technologies Monitor*, April 26, 2013, http://ghgnews.com/index.cfm/miss-power-to-absorb-540m-in-cost-increasesfrom-kemper-plant/. Other reports cite the total costs for the plant.

[41] See Mississippi Power, Kemper IGCC Project, Monthly Status Report Through December 2013, http://www.sec.gov/ Archives/edgar/data/66904/000009212214000002/msmonthly report8-kex99x01.htm.

[42] See, for example, Mark Drajem, "Mississippi's Kemper Coal Plant Overruns Show Risk of EPA Carbon Rule," *Bloomberg News, SunHerald.com*, September 19, 2013, http://www.sunherald.com/2013/09/19/4964367/kempercounty-coal-plant-overruns.html.

[43] U.S. Department of Energy, *Strategic Plan*, May 2011, p. 18, http://energy.gov/sites/prod/files/ 2011_DOE_Strategic_Plan_.pdf.

[44] See, for example, the presentations given by Edward Rubin of Carnegie Mellon University, Howard Herzog of the Massachusetts Institute of Technology, and Jeff Phillips of the Electric Power Research Institute, at the CRS seminar *Capturing Carbon for Climate Control: What's in the Toolbox and What's Missing*, November 18, 2009. (Presentations available from Peter Folger at 7-1517.) Rubin stated that at least 10 full-scale demonstration projects would be needed to establish the reliability and true cost of CCS in power plant applications. Herzog also called for at least 10 demonstration plants worldwide that capture and sequester a million metric tons of CO2 per year. In his presentation, Phillips stated that large-scale demonstrations are critical to building confidence among power plant owners.

[45] DOE had solicited and awarded funding for CCPI projects in two previous rounds of funding: CCPI Round 1 and Round 2. The Recovery Act funds were to be allocated CCPI Round 3, focusing on projects that utilize CCS technology and/or the beneficial reuse of CO2. For more details, see http://www.fossil.energy.gov/programs/ powersystems/cleancoal/.

[46] U.S. Department of Energy, *Detailed Budget Justifications FY2010*, volume 7, Fossil Energy Research and Development, p. 35, http://www.cfo.doe.gov/budget/10budget/Content /Volumes/Volume7.pdf.

[47] DOE 2010 *CCS Roadmap*, p. 15.

[48] The first solicitation closing date was January 20, 2009; the second solicitation closing date was August 24, 2009. Thus the first set of project proposals were submitted prior to enactment of the Recovery Act. See http://www.fossil.energy.gov/programs/powersystems /cleancoal/.

[49] MIT Carbon Capture & Sequestration Technologies, *Plant Barry Fact Sheet: Carbon Dioxide Capture and Storage Project*, http://sequestration.mit.edu/tools/projects/plant_barry.html.

[50] Ibid.

[51] U.S. DOE, Fossil Energy Techline, *Secretary Chu Announces Two New Projects to Reduce Emissions from Coal Plants*, July 1, 2009, http://www.fossil.energy.gov/news/techlines /2009/09043-DOE_Announces_CCPI_Projects.html.

[52] Lauren Donovan, "Basin Shelves Lignite's First Carbon Capture Project," *Bismarck Tribune*, December 17, 2010, http://bismarcktribune.com/news/local/a5fb7ed8-0a1b-11e0-b0ea-001cc4c03286.html.

[53] Daryl Hill and Tracie Bettenhausen, "Fresh Tech, Difficult Decisions: Basin Electric has a History of Trying New Technology," Basin Electric Power Cooperative newsletter, January-February 2011, http://www.basinelectric.com/ Miscellaneous/pdf/FeatureArticles /Fresh_Tech,_difficul.pdf.

[54] U.S. DOE, Fossil Energy Techline, *Secretary Chu Announces $3 Billion Investment for Carbon Capture and Sequestration*, December 4, 2009, http://www.fossil.energy.gov /news/techlines/2009/09081- Secretary_Chu_Announces_CCS_Invest.html.

[55] Matthew L. Wald and John M. Broder, "Utility Shelves Ambitious Plan to Limit Carbon," *New York Times*, July 13, 2011, http://www.nytimes.com/2011/07/14/business/energy-environment/utility-shelves-plan-to-capture-carbondioxide.html?_r=1.

[56] Michael G. Morris, chairman of AEP, quoted in Matthew L. Wald and John M. Broder, "Utility Shelves Ambitious Plan to Limit Carbon," *The New York Times*, July 13, 2011.

[57] Wald and Broder, *New York Times*, July 13, 2011.

[58] MIT Carbon Capture & Sequestration Technologies, *AEP Mountaineer Fact Sheet: Carbon Dioxide Capture and Storage Project*, http://sequestration.mit.edu/tools/projects /aep_ alstom_mountaineer.html.

[59] Telephone conversation with Joseph Giove, DOE Office of Fossil Energy, March 19, 2012.

[60] U.S. DOE Fossil Energy Techline, "Secretary Chu Announces Up To $154 Million for NRG Energy's Carbon Capture and Storage Project in Texas," March 9, 2010, http://www.fossil.energy.gov/news/techlines/2010/10005-NRG_Energy_Selected_to_Receive_DOE.html.

[61] Telephone conversation with Joseph Giove, DOE Office of Fossil Energy, April 11, 2011.

[62] U.S. Congress, House Committee on Appropriations, Subcommittee on Military Construction, Veterans Affairs, and Related Agencies, *Military Construction and Veterans Affairs and Related Agencies Appropriations Act, 2012*, conference report to accompany H.R. 2055, 112th Cong., 1st sess., December 15, 2011, H.Rept. 112-331 (Washington: GPO, 2011), p. 851.

[63] Telephone conversation with Joseph Giove, DOE Office of Fossil Energy, March 19, 2012.

[64] Email from Regis K. Conrad, Director, Division of Cross-Cutting Research, DOE, March 20, 2012.

[65] U.S. DOE, National Energy Technology Laboratory, *Major Demonstrations, Industrial Capture and Storage (ICCS): Area 1*, http://www.netl.doe.gov/technologies /coalpower/cctc/iccs1/index.html#.

[66] U.S. DOE, *Recovery Act, Innovative Concepts for Beneficial Reuse of Carbon Dioxide*, http://fossil.energy.gov/ recovery/projects/beneficial_reuse.html.

[67] Email from Regis K. Conrad, Director, Division of Cross-Cutting Research, DOE, March 20, 2012.

[68] U.S. DOE, Carbon Capture and Storage from Industrial Sources, Industrial Carbon Capture Project Selections, http://fossil.energy.gov/recovery/projects/iccs_projects_0907101.pdf.

[69] See http://www.fossil.energy.gov/news/techlines/2008/08003-DOE_Announces_Restructured_ FutureG.html.

[70] Prior to when DOE first announced it would restructure the program in 2008, the FutureGen Alliance announced on December 18, 2007, that it had selected Mattoon, IL, as the host site from a set of four finalists. The four were Mattoon, IL; Tuscola, IL; Heart of Brazos (near Jewett, TX); and Odessa, TX.

[71] See DOE announcement on June 12, 2009, http://www.fossil.energy.gov/news /tcchlines/2009/09037- DOE_Announces_FutureGen_Agreement.html.

[72] See DOE Techline, http://www.netl.doe.gov/publications/press/2010/10033- Secretary_Chu_ Announces_FutureGen_.html.

[73] Ameren had planned to replace the oil-fired boiler with a coal-fired boiler using oxy-combustion technology to allow carbon capture. See http://www.futuregenalliance.org /pdf/FutureGen%20FAQ-General%20042711.pdf.

[74] For more information about the history of FutureGen, and issues for Congress, see CRS Report R43028, *The FutureGen Carbon Capture and Sequestration Project: A Brief History and Issues for Congress*, by Peter Folger

[75] For more information on FutureGen, see CRS Report R43028, *The FutureGen Carbon Capture and Sequestration Project: A Brief History and Issues for Congress*, by Peter Folger

[76] E. S. Rubin, "The Government Role in Technology Innovation: Lessons for the Climate Change Policy Agenda," Institute of Transportation Studies, 10th Biennial Conference on Transportation Energy and Environmental Policy, University of California, Davis, CA (August 2005).

[77] National Aeronautics and Space Administration, "Definition of Technology Readiness Levels," at http://esto.nasa.gov/files/TRL_definitions.pdf.

[78] For a more thorough discussion of different schemes describing stages of technology development, see chapter 4 of CRS Report R41325, *Carbon Capture: A Technology Assessment*, by Peter Folger.

[79] E. S. Rubin, "The Government Role in Technology Innovation: Lessons for the Climate Change Policy Agenda," Institute of Transportation Studies, 10th Biennial Conference on Transportation Energy and Environmental Policy, University of California, Davis, CA (August 2005).

[80] Another possible source of uncertainty for FutureGen, and other large industrial CCS projects, is cost recovery during the operating phase of the plant after the construction phase and initial capital investments are made. "Learning by doing" should increase operating efficiency, but it is unclear by how much and over what time span. For more discussion on

cost trajectories and expected efficiency gains, see CRS Report R41325, *Carbon Capture: A Technology Assessment*, by Peter Folger.

[81] The total DOE share for the 10 projects is $46.6 million. See DOE, *Recovery Act*, http://fossil.energy.gov/recovery/ projects/site_characterization.html.

[82] DOE 2010 *CCS Roadmap*, p. 55.

[83] Four Canadian provinces are partners with DOE in two of the regional partnerships, and are members with other participating organizations that are contributing funding and other support to the partnerships.

[84] DOE National Energy Technology Laboratory, Carbon Sequestration Regional Carbon Sequestration Partnerships, http://www.netl.doe.gov/technologies/carbon_seq/infrastructure /rcsp.html.

[85] Ibid.

[86] For details on the two large-scale injection experiments by SECARB, see http://www.secarbon.org/; for details on the large-scale injection experiment by MGSC, see http://sequestration.org/.

[87] DOE National Energy Technology Laboratory, *Carbon Sequestration Regional Partnership Development Phase (Phase III) Projects*, http://www.netl.doe.gov/technologies /carbon_seq/infrastructure/rcspiii.html.

[88] For more information on the EPA Class VI wells in particular, and the Safe Drinking Water Act generally, see CRS Report RL34201, *Safe Drinking Water Act (SDWA): Selected Regulatory and Legislative Issues*, by Mary Tiemann.

[89] For a discussion of several of these legal issues, see CRS Report RL34307, *Legal Issues Associated with the Development of Carbon Dioxide Sequestration Technology*, by Adam Vann and Paul W. Parfomak.

[90] For more information on the different issues regarding community acceptance of CCS, see CRS Report RL34601, *Community Acceptance of Carbon Capture and Sequestration Infrastructure: Siting Challenges*, by Paul W. Parfomak.

[91] Testimony of Scott Klara, Deputy Laboratory Director, National Energy Technology Laboratory, U.S. Department of Energy, in U.S. Congress, Senate Energy and Natural Resources Committee, *Carbon Capture and Sequestration Legislation*, hearing to receive testimony on carbon capture and sequestration legislation, including S. 699 and S. 757, 112th Cong., 1st sess., May 12, 2011, S.Hrg. 112-22.

In: Deployment of Carbon Capture and ... ISBN: 978-1-63117-726-2
Editor: Denis Tiernan

Chapter 3

The FutureGen Carbon Capture and Sequestration Project: A Brief History and Issues for Congress*

Peter Folger

Summary

More than a decade after the George W. Bush Administration announced its signature clean coal power initiative—FutureGen—the program is still in early development. Since its inception in 2003, FutureGen has undergone changes in scope and design. As initially conceived, FutureGen would have been the world's first coal-fired power plant to integrate carbon capture and sequestration (CCS) with integrated gasification combined cycle (IGCC) technologies. FutureGen would have captured and stored carbon dioxide (CO_2) emissions from coal combustion in deep underground saline formations and produced hydrogen for electricity generation and fuel cell research. Increasing costs of development, among other considerations, caused the Bush Administration to discontinue the project in 2008. In 2010, under the Obama Administration, the project was restructured as FutureGen 2.0: a coal-fired power plant that would integrate oxycombustion technology to capture CO_2. FutureGen 2.0 is the U.S. Department of Energy's (DOE)

* This is an edited, reformatted and augmented version of a Congressional Research Service publication, CRS Report for Congress R43028, prepared for Members and Committees of Congress, from www.crs.gov, dated February 10, 2014.

most comprehensive CCS demonstration project, combining all three aspects of CCS technology: capturing and separating CO_2 from other gases, compressing and transporting CO_2 to the sequestration site, and injecting CO_2 in geologic formations for permanent storage.

Congressional interest in CCS technology centers on balancing the competing national interests of fostering low-cost, domestic sources of energy like coal against mitigating the effects of CO_2 emissions in the atmosphere. FutureGen 2.0 would address these interests by demonstrating CCS technology as commercially viable. Among the challenges to the development of FutureGen 2.0 are rising costs of production, ongoing issues with project development, lack of incentives for investment from the private sector, and time constraints. Further, FutureGen's development would need to include securing private sector funding to meet increasing costs, purchasing the power plant for the project, obtaining permission from DOE to retrofit the plant, performing the retrofit, and then meeting the goal of 90% capture of CO_2.

The FutureGen project was conceived as a public-private partnership between industry and DOE with agreements for cost-share and cooperation on development, demonstration, and deployment of CCS technology. The public-private partnership has been criticized for leading to setbacks in FutureGen's development, since the private sector lacks incentives to invest in costly CCS technology. Regulations, tax credits, or policies such as carbon taxation or cap-and-trade that increase the price of electricity from conventional power plants may be necessary to make CCS technology competitive enough for private sector investment. Even then, industry may choose to forgo coal-fired plants for other sources of energy that emit less CO_2, such as natural gas. However, Congress signaled its support for FutureGen 2.0 via the American Recovery and Reinvestment Act of 2009 (ARRA, P.L. 111-5) by appropriating almost $1 billion for the project. ARRA funding will expire on September 30, 2015, and it remains a question whether the project will expend all of its federal funding before that deadline.

A proposed rule by the Environmental Protection Agency (EPA) to limit CO_2 emissions from new fossil-fuel power plants may provide some incentive for industry to invest in CCS technology. The debate has been mixed as to whether the rule would spur development and deployment of CCS for new coal-fired power plants or have the opposite effect. Multiple analyses indicate that there will be retirements of U.S. coal-fired capacity; however, virtually all analyses agree that coal will continue to play a substantial role in electricity generation for decades. The rapid increase in the domestic natural gas supply as an alternative to coal, in combination with regulations that curtail CO_2 emissions, may lead electricity producers to invest in natural gas-fired plants, which emit approximately half the amount of CO_2 per unit of electricity produced compared to coal-fired plants.

INTRODUCTION AND BACKGROUND

This report briefly summarizes the history of FutureGen, discusses why it has gained interest and support from some Members of Congress and the Administration while remaining in initial stages of development, and offers some policy considerations on barriers that challenge its further development as a model for a CCS program. A timeline history of FutureGen is found at the end of this report.

FutureGen is a clean-coal technology program managed through a public-private partnership between the U.S. Department of Energy (DOE) and the FutureGen 2.0 Industrial Alliance. The FutureGen program as originally conceived in 2003 by the George W. Bush Administration had the intent of constructing a net zero-emission fossil-fueled power plant with carbon capture and sequestration (CCS) technology.[1] CCS is a process envisioned to capture carbon dioxide (CO_2)— a greenhouse gas associated with climate change— emitted from burning fossil fuels and store it in deep underground geologic formations, thus preventing its release into the atmosphere. If widely deployed in the United States, CCS could decrease the amount of U.S.-emitted CO_2. In 2008, DOE withdrew from the FutureGen partnership, citing rising costs of construction as its reason. Subsequently, DOE restructured the FutureGen program to instead develop two or three demonstration projects at different power plants around the country. In 2010, the Obama administration announced another change to the program with the introduction of FutureGen 2.0, which would retrofit an existing fossil fuel power plant in Illinois with CCS technology.[2]

The FutureGen project was originally conceived as a cost-share between the federal government, which would cover 76% of the cost, and the private sector, which would provide the remaining 24%. Between FY2004 and FY2008, Congress appropriated $174 million to the original FutureGen project. DOE obligated $44 million and expended $42 million between FY2005 and FY2010 toward the original project.[3] Under the Obama Administration, Congress appropriated almost $1 billion in the American Recovery and Reinvestment Act of 2009 (ARRA, P.L. 111-5) for FutureGen 2.0. Furthermore, DOE has obligated nearly $60 million but has expended $2 million from regular appropriations to FutureGen 2.0 since FY2010.[4] Together with the approximately $74 million expended on the project from ARRA funding (discussed below), total expenditures for FutureGen since its conception were between $110 and $120 million as of early 2014.

The FutureGen Industrial Alliance estimated the total cost of the FutureGen 2.0 program to be nearly $1.3 billion, with $730 million used toward retrofitting and repowering Ameren Corporation's power plant in Meredosia, Illinois, and $550 million used for the construction of a CO_2 pipeline, storage site, and training and research center. In 2011, they estimated that the project would create approximately 1,000 construction jobs and another 1,000 jobs for suppliers across the state.[5] A 2013 report from the University of Illinois predicted that the project could create an average of 620 permanent jobs for 20 years and approximately $12 billion of business volume by 2037 for the state of Illinois.[6]

DOE CCS Programs

Current scientific research associates an increase in atmospheric GHGs (in particular CO_2, methane, and nitrous oxides), which trap heat in the earth's atmosphere, with the potential for changing the Earth's climate. The increase in the atmospheric concentration of CO_2 in the 20th and 21st centuries is due almost entirely to human activities.[7] If successful, FutureGen 2.0 would demonstrate a technology that, if widely deployed, could capture a significant fraction of U.S. CO_2 emissions for geologic sequestration.

DOE's Office of Fossil Energy directs three major CCS programs: the Clean Coal Power Initiative (CCPI), Industrial Carbon Capture and Storage (ICCS), and FutureGen 2.0.[8] Through its CCPI program, DOE partners with industry leaders in a cost-share arrangement to develop new CCS technologies for power plant utilities in order to reduce greenhouse gas emissions by boosting plant efficiencies and capturing CO_2. Of the six projects selected under the most recent funding for CCPI (Round 3), three have withdrawn, citing concerns over costs and regulations. DOE's share for the three active projects is $1.0 billion of a total $6.1 billion, approximately 17%. DOE is also partnering with industry for 31 projects in the ICCS program, which supports R&D in a non-utility large-scale industrial CCS program and a program to support beneficial CO_2 use. The combined total DOE share for all the ICCS projects is $1.422 billion of a total $2.0 billion, approximately 70%.

FutureGen 2.0 is DOE's most comprehensive CCS demonstration project, combining all three aspects of CCS technology: capturing and separating CO_2 from other gases, compressing and transporting CO_2 to the sequestration site, and injecting CO_2 in geologic formations.

Current Status of FutureGen

In October 2010, FutureGen 2.0 developers began working on Phase 1 of the project with the Pre-Front End Engineering Design (Pre-FEED) report, which included plant design, estimated project cost, and basis for applying for National Environmental Policy Act (NEPA) review and other state and local permits.[9] The report showed that the estimated price for FutureGen 2.0 had increased from $1.3 billion to $1.65 billion. Subsequently, cost reduction measures were identified and implemented, including establishing the plant gross output at 168 MW (the steam turbine is nominally rated at 200 MW), and using a combination of 60% Illinois coal and 40% Powder River Basin (PRB) coal to reduce sulfur and chlorine emissions.[10] Also, in late 2011 Ameren announced it was closing its power plant in Meredosia, Illinois, and discontinuing its cooperative agreement with DOE.[11] Following that announcement, the project was redesigned to reflect that the Alliance would take control of the capture process as well as the transportation and storage site. The Alliance is currently negotiating the purchase of parts of the Meredosia Energy Center from Ameren to continue with project development. *Figure 1* shows the location of the town of Meredosia, Illinois, the proposed pipeline route, and the proposed carbon sequestration site where the captured CO_2 would be injected underground and stored. Throughout the summer and fall of 2012, the project continued to confront rising cost estimates, as well as challenges in negotiating a long-term power purchasing agreement with the state of Illinois.[12] However, the project has achieved several milestones since 2012 that could favor its future progress. In late December 2012, the Illinois Commerce Commission voted 3-2 to approve a power procurement plan for the state that requires utilities to purchase all the electricity generated by the FutureGen 2.0 facility for 20 years. That decision cleared a major hurdle for FutureGen 2.0, and the decision allows Commonwealth Edison and Ameren Illinois to collect costs for the project from the state's alternative retail electric suppliers.[13] Opposition to the power procurement proposal stemmed primarily from those opposed to its potential to raise costs for retail customers.[14]

In February 2013, DOE approved the start of Phase 2 of the project, which includes final permitting and design activities that precede a decision to begin construction.[15] The project faced delays while it was being redesigned following the release of the Pre-FEED report; however, the FEED activities resulted in a 70%-90% design completion for the project, which is better than the industry standard of about 25%, according to the FutureGen 2.0 Industrial Alliance.[16]

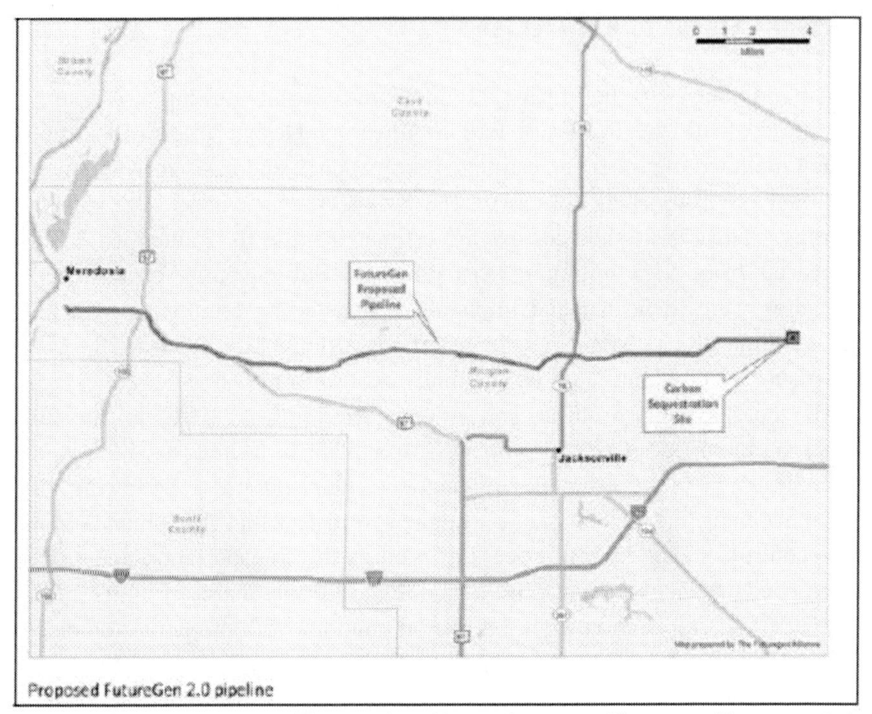

Source: The FutureGen Alliance, http://www.futuregenalliance.org/futuregen-2-0-project

Notes: The proposed pipeline is approximately 30 miles long. Construction is anticipated to begin in the summer of 2014, according to the FutureGen 2.0 Industrial Alliance.

Figure 1. Map Showing the Town of Meredosia, IL, the Proposed Pipeline Route, and the Proposed CO_2 Sequestration Site.

On October 25, 2013, DOE issued the final environmental impact statement (EIS) for FutureGen 2.0. The proposed action in the EIS is for DOE to provide funding of approximately $1 billion to the FutureGen 2.0 Industrial Alliance to support the completion of Phase 2—preliminary and final design for the project—followed by construction and commissioning (Phase 3) and operations (Phase 4).[17] On November 30, 2013, EPA published a notice of availability in the *Federal Register*, [18] and on January 16, 2014, DOE issued a favorable record of decision (ROD), as part of the NEPA process. Issuance of the ROD clears the last hurdle in the NEPA process, and reportedly allows the FutureGen 2.0 Industrial Alliance to move forward pending approval of a permit to install the CO_2 injection wells and meeting financial requirements."[19]

POLICY CHALLENGES AND ISSUES FOR CONGRESS

After more than 10 years and two restructuring efforts since FutureGen's inception, the project is still in its early development stages. Although the FutureGen 2.0 Industrial Alliance completed drilling a characterization well at the storage site in Morgan County, IL, and installed a service rig over the well for further geologic analysis, issues with the power plant itself have not yet been resolved. Among the remaining challenges are securing private sector funding to meet increasing costs, purchasing the Meredosia power plant from Ameren, obtaining permission from the DOE to retrofit the plant, performing the retrofit, and then meeting the goal of 90% capture of CO_2. In addition, the Alliance is awaiting approval for a Class VI well permit for the injection and sequestration wells.[20]

Cost, Schedule, and Funding

Project Costs

Increasing projected costs have posed significant problems for FutureGen's development since 2003. Confronted with increasing projected costs in 2008, DOE under the George W. Bush Administration first restructured FutureGen, then postponed the program when cost projections rose from $950 million to $1.8 billion.

When Secretary of Energy Steven Chu announced the new FutureGen 2.0 in 2010, the cost was estimated at $1.3 billion, with the DOE covering 80% of costs and industry partners contributing the remaining 20% of the total. Initially, FutureGen 2.0 was to be implemented through two separate cooperative agreements, with approximately $590 million of ARRA funds allocated to Ameren Corporation to retrofit a power plant[21] and approximately $459 million of ARRA funds to the FutureGen 2.0 Industrial Alliance to implement a pipeline and regional CO_2 storage reservoir project.[22]

According to the FutureGen 2.0 Industrial Alliance, total capital costs for the FutureGen 2.0 project are estimated to be $1.65 billion.[23] The Alliance is expected to cover the additional cost beyond the original cost estimate for FutureGen 2.0. Rising costs of construction may continue to be a challenge to the project's development.

Schedule and Funding

Some projections for FutureGen predict construction on the power plant, pipeline, and storage facility will conclude by 2017.[24] A looming question is whether the FutureGen 2.0 Alliance will have sufficient time to expend the nearly $995 million of ARRA funding appropriated by Congress for the project before it expires on September 30, 2015. As of October 2013, the FutureGen 2.0 Alliance has expended about $73.97 million, or about 7.4%, of the total $994,729,000 appropriated under ARRA.[25]

Once construction begins, the rate of spending will undoubtedly increase. Now that the DOE ROD has been issued, it is likely that construction will begin sometime in spring or early summer of 2014. But even if construction began as early as March 2014, the project would need to spend approximately $921 million over 19 months, or about $48 million per month until the end of September 2015 to exhaust all the ARRA funding. According to the investigatory work of one industry observer, using documents obtained from DOE under a Freedom of Information Act request, DOE would grant the Alliance the flexibility to accelerate the cost-share and expend the ARRA-provided funding to cover capital costs before using private funds from the Alliance to cover its portion of the cost-share.[26] According to the report, DOE would require an increased level of oversight over the project to safeguard the public investment in the project. Further, DOE would have the ability to suspend or terminate funding if the project failed to demonstrate sufficient progress.[27]

Public-Private Partnership

The partnership between the federal government and the private sector in funding and developing FutureGen has been marked by a series of setbacks and challenges. Some critics of the public-private partnership attribute the project's decade-long stasis to a lack of incentives for industry leaders to invest seriously in clean coal technologies. A report released by the Massachusetts Institute of Technology in 2007 stated that government investment and leadership in carbon capture technologies are necessary: "Given the technical uncertainty and the current absence of a carbon charge, there is no economic incentive for private firms to undertake such projects."[28] Since the MIT report was published, Congress has appropriated nearly $7 billion in CCS research and development (R&D), including FutureGen;

however, Congress has not enacted any form of a "carbon charge," through either a cap-and-trade system or a carbon tax.[29]

Ameren Corporation, which partnered with DOE to retrofit its power plant in Meredosia, IL, for FutureGen 2.0, discontinued operations at the Meredosia Energy Center where the plant is located, stating that it would not be able to afford the costs to comply with air pollution rules issued in July 2011 by the EPA to reduce sulfur dioxide and nitrogen oxide.[30] In addition to the FutureGen project, DOE partnered with industry for six other commercial-scale CCS projects through its Clean Coal Power Initiative (CCPI) program.[31] The 2010 DOE Strategic Plan report predicted that at least five of DOE's major CCS projects would become operational by 2016.[32] Since the report was released, three of the six industry partners of CCPI projects have pulled out of agreements with DOE. The departure of several industry leaders from contracts with DOE demonstrates the volatility of the public-private partnership model.

EPA Proposed Rule to Limit CO_2 from New Power Plants

On September 20, 2013, the U.S. Environmental Protection Agency (EPA) re-proposed a standard that would limit emissions of carbon dioxide (CO_2) from new fossil-fueled power plants. As re-proposed, the rule would limit emissions to no more than 1,100 pounds per megawatt-hour of production from new coal-fired power plants and between 1,000 and 1,100 pounds per megawatt-hour (depending on size of the plant) for new natural gas-fired plants. EPA proposed the standard under Section 111 of the Clean Air Act. According to EPA, new natural gas-fired stationary power plants should be able to meet the proposed standard without additional cost and the need for add-on control technology. However, new coal-fired plants would be able to meet the standard only by installing carbon capture and sequestration (CCS) technology to capture about 40% of the CO_2 they typically produce. The proposed standard allows for a seven-year compliance period for coal-fired plants but would demand a more stringent standard for those plants that limits CO_2 emissions to an average of 1,000-1,050 pounds per megawatt-hour.[33]

On January 8, 2014, EPA published the re-proposed rule in the *Federal Register*.[34] Publishing in the *Federal Register* triggers the start of a 60-day public comment period: comments will be accepted until March 10, 2014. The initial 2012 proposal generated more than 2.5 million comments, which prompted, in part, the September 20, 2013, re-proposal. Promulgation of the final rule could be expected sometime after the public comment period ends and EPA evaluates the comments.

The re-proposed rule would address only new power plants. However, Section 111 of the Clean Air Act requires that EPA develop standards for greenhouse gas emissions for existing plants whenever it promulgates standards for new power plants. In his June 25, 2013, memorandum, President Obama directed the EPA to issue proposed standards for existing plants by June 1, 2014, and to issue final rules a year later.

Given the pending EPA rule, congressional interest in the future of coal as a domestic energy source appears directly linked to the future of CCS. The history of CCS research, development, and deployment (RD&D) at DOE and the pathway of its signature program—FutureGen—invite questions about whether DOE-funded RD&D will enable widespread deployment of CCS in the United States within the next decade.

The Natural Gas Alternative?

When EPA first proposed a new rule regulating GHG emissions from power plants that would likely require CCS, Congress considered legislation to block the new regulations. For example, the Subcommittee on Energy and Power of the House Science, Space, and Technology Committee held a hearing on June 19, 2012, where opponents of the new rule, including FutureGen Alliance Chairman Steven E. Winberg, criticized the regulations: "In effect, EPA's rule will eliminate any new coal for years to come because EPA is requiring new coal-fueled power plants to meet a natural gas equivalent CO_2 standard, before CCS technology is commercially available."[35]

Following the September 20, 2013, re-proposal of the rule, the debate has been mixed as to whether the rule would spur development and deployment of CCS for new coal-fired power plants or have the opposite effect. Multiple analyses indicate that there will be retirements of U.S. coal-fired capacity; however, virtually all analyses agree that coal will continue to play a substantial role in electricity generation for decades. How many retirements would take place and the role of EPA regulations in causing them are matters of dispute.[36]

Since the September 2013 re-proposal, the argument over the rule has focused, in part, on whether CCS is the best system of emissions reduction (BSER) for coal plants and whether it has been "adequately demonstrated" as such, as required under the Clean Air Act. In its re-proposed rule, EPA cites the "existence and apparent ongoing viability" of several ongoing CCS demonstration projects as examples that justify a separate determination of BSER for coal-fired plants and integrated gasification combined-cycle plants. (The second BSER determination is for gas-fired power plants.)[37] The EPA noted that these projects had reached advanced stages of construction and development, "which suggests that proposing a separate standard for coal-fired units is appropriate." FutureGen 2.0 was not included as one of the projects used to justify the proposed rule, despite its 10-year long history and more than $1 billion in committed federal support. Its omission from the EPA re-proposed rule further reinforces FutureGen's status as a CCS project in the early stages of development.

The huge increase in the U.S. domestic supply of natural gas, due largely to the exploitation of unconventional shale gas reservoirs through the use of hydraulic fracturing, has also led to a shift to natural gas for electricity production.[38] The shift appears to be largely due to the cheaper and increasingly abundant fuel—natural gas—compared to coal for electricity production. The EPA re-proposed rule, discussed above, noted that "power companies often choose the lowest cost form of generation when determining what type of new generation to build. Based on [Energy Information Administration] modeling and utility [Integrated Resource Plans], there appears to be a general acceptance that the lowest cost form of new power generation is [natural gas combined-cycle]." Cheap gas, due to the rapid increase in the domestic natural gas supply as an alternative to coal, in combination with regulations that curtail CO_2 emissions, may lead electricity producers to invest in natural gas-fired plants, which emit approximately half the amount of CO_2 per unit of electricity produced compared to coal-fired plants. Regulations and abundant cheap gas may raise questions about the rationale for CCS demonstration projects like FutureGen.

Alternatively, and despite increasingly abundant domestic natural gas supplies, EPA regulations could provide the necessary incentives for the industry to accelerate CCS development and deployment for coal-fired power plants. As part of its re-proposed ruling, EPA cites technology as one of four factors that it considers in making a BSER determination.[39] Specifically, EPA stated that it "considers whether the system promotes the implementation and further development of technology," in this case referring to CCS technology. It appears that EPA asserts that its rule would likely promote CCS development and deployment rather than hinder it. Those arguing against the re-proposed rule do so on the basis that CCS technology has not been adequately demonstrated, and that it violates provisions in P.L. 109-58, the Energy Policy Act of 2005, that prohibit EPA from setting a performance standard based on the use of technology from certain DOE-funded projects, such as the three projects cited in the EPA re-proposal, among other reasons.[40]

On January 9, 2014, Representative Whitfield and 62 cosponsors introduced H.R. 3826, the Electricity Security and Affordability Act, which would essentially impose a number of requirements to be met before EPA could issue greenhouse gas emission regulations under Section 111 of the Clean Air Act, such as the EPA re-proposed rule discussed above. On January 14, 2014, the Energy and Power subcommittee, House Energy and Commerce committee, voted to report the bill. Much of the discussion during the bill's markup centered on whether CCS was an adequately demonstrated technology to meet the requirements of the Clean Air Act.

OUTLOOK

Congressional consideration of CCS has focused on balancing competing national interests, such as fostering low-cost domestic sources of energy like coal versus reducing greenhouse gas (GHG) emissions in the atmosphere. Legislative proposals during the 109th and 110th Congresses focused on advancing carbon capture technologies that reduce CO_2 emissions to mitigate GHG-induced global warming. Congress began appropriating funds specifically for FutureGen beginning in 2005. Previously, DOE had allocated funds under its Clean Coal Power Initiative (CCPI) program. With the American Recovery and Reinvestment Act of 2009, Congress appropriated approximately $1 billion for the FutureGen 2.0 project.

The revival of FutureGen under the Obama Administration as FutureGen 2.0 has sparked increased scrutiny of the future of integrated CCS technology on a commercially viable scale. FutureGen was originally proposed to demonstrate the feasibility of using CCS technology to mitigate CO_2 emissions into the atmosphere. Among the challenges that continue to influence the development of FutureGen 2.0 are rising costs of construction, ongoing issues with project development, lack of incentives for investment from the private sector, and time constraints on project development. Despite congressional and Obama Administration commitments to the FutureGen 2.0 project, particularly the $1.0 billion appropriation from ARRA, questions remain as to whether or not FutureGen 2.0 will succeed.

The Congressional Budget Office (CBO) published a report in June 2012 stating that the success of CCS technology depends on reducing technical costs, ensuring the effectiveness of CCS, and adopting policies that provide incentives for industry to pursue the high-cost demonstration technologies.[41] The report explained that if regulations, tax credits, or policies such as carbon taxation or cap-and-trade that increase the price of electricity from conventional power plants are adopted, then CCS technology may become competitive enough for private sector investment. Even then, industry may choose to forgo coal-fueled plants for natural gas or other sources that emit less CO_2 compared to coal, according to CBO.[42]

TIMELINE

The timeline that follows shows a chronology of the history of FutureGen since 2003.

PHASE 1

2003

February 27: President George W. Bush proposed a 10 year, $1 billion project to build a 275 MW coal-fired power plant that would integrate carbon sequestration and hydrogen production.

2004

March: In its 2004 Report to Congress, the DOE estimated that FutureGen would cost $950 million with the DOE contributing 76% and the private sector the remaining 24% of the total cost.

July: The FutureGen Industrial Alliance, a non-profit company composed of the largest international coal companies and electric utilities, was formed to partner with the DOE on the development of FutureGen. The seven founding Alliance members are American Electric Power, BHP Billiton, CONSOL Energy Inc., Foundation Coal Corporation, Kennecott Energy Company (a member of the Rio Tinto group), Peabody Energy, and Southern Company.

2005

October 27: China Huaneng Group, China's largest coal-fueled power generator, joined the Alliance.

December: The DOE and the Alliance signed a Cooperative Agreement partnering in all development aspects of the $1 billion FutureGen project, including site and technology selection, construction and operation.

2006

February 23: Anglo American, one of the world's largest diversified mining and natural resource groups, joined the FutureGen Industrial Alliance as its ninth member.

March 8: The Alliance released the final Request for Proposals (RFP) for regions interested in hosting the world's first coal-fueled "zero emissions" power plant.

May 23: PPL Corporation, an electric company delivering electricity and natural gas in the United States and United Kingdom, joined the FutureGen Industrial Alliance as its tenth member.

July 25: The Alliance selected four finalist hosting sites for FutureGen: Mattoon, IL, Tuscola, IL, Odessa, TX and Jewett, TX.

October 31: E.ON US, the world's largest investor-owned electric utility service provider, joined the Alliance as its eleventh member.

December 7: Xstrata Coal, Australia-based exporter of high energy thermal coal, joined the Alliance as its twelfth member.

December 15: The United States and China announced their cooperation on FutureGen and signed an Energy Efficiency Protocol during the first U.S.-China Strategic and Economic Dialogue.

2007

January: The Alliance produced an initial conceptual design report for the original FutureGen project estimating the cost of the program at $1.8 billion accounting for inflation through 2017.

March 23: The DOE and Alliance signed a Cooperative Agreement stipulating that the DOE would cover 74% and the Alliance would share the remaining 26% of the $1.8 billion cost.

May 25: The DOE released a Draft Environmental Impact Statement (DEIS) that included a review of all four candidate sites in Illinois and Texas.

November 9: The DOE released a final Environmental Impact Statement (EIS) predicting program costs at $1.8 billion with projected revenues from the sale of electricity at $301 million.

December: Given rising costs of FutureGen development, DOE's Office of Fossil Energy attempted to negotiate a new cost-sharing arrangement with the Alliance before continuing the cooperative agreement in June 2008.

December 6: Luminant, a Texas-based electric utility, joined the Alliance as the thirteenth member.

December 10: The DOE advised the Alliance not to announce the selected plant site.

December 11: Peabody Energy, a founding member of the Alliance, became an equity partner in China's $1 billion GreenGen project, an envisioned near-zero emission carbon capture project modeled after FutureGen. Peabody has a 6% stake in GreenGen.

December 18: The Alliance announced its selection of Mattoon, IL as the final site to host the FutureGen power plant.

2008

January 8: Alliance board of directors elected Paul W. Thompson of E.ON U.S. as its next chairman of the board to replace outgoing chairman Greg A. Walker.

PHASE 2

April 9: Alliance CEO Michael J. Mudd told the Senate Subcommittee on Science, Technology and Innovation that DOE's recent proposal to restructure FutureGen failed to address the challenges of climate change and energy security and would delay CCS technology by several years.

April 15: Alliance Chairman Paul Thompson testified before the House Science and Technology Committee that costs of all global energy infrastructure projects increased due to inflation. Thompson said that FutureGen costs were consistent with industry average increases.

May 7: The DOE released a draft Funding Opportunity Announcement for restructured FutureGen to receive public input and gauge public interest in the project.

May 8: Secretary of Energy Samuel Bodman testified before the Senate Subcommittee on Energy and Water Development that the cost of the FutureGen project doubled from $950 million to $1.8 billion. Alliance Chairman Paul Thompson told members of the subcommittee to continue supporting the original FutureGen project in Mattoon, IL because of predicted delays and reduced standards of CO2 capture in DOE's restructured FutureGen project. The Senate subcommittee held the oversight hearing to discuss DOE's decision to restructure FutureGen.

May 19: Senators Kit Bond (R-MO) and Dick Durbin (D-IL) sent a letter to Secretary of Energy Samuel Bodman to extend the budget period of the existing cooperative agreement from June 15, 2008 to March 30, 2009 in order to retain funds already appropriated for FutureGen, maintain the original FutureGen program and allow the incoming administration to make a decision on the future of FutureGen.

June: Senior DOE officials directed the Office of Fossil Energy to negotiate a new cost-sharing agreement with the Alliance under the Cooperative Agreement that was scheduled for a continuation in June. The negotiations failed to yield an agreement.

June: The DOE formally discontinued its cost-share with the Alliance for FutureGen. Luminant and PPL Corporation pulled out of the FutureGen Alliance.

July: The Senate Energy and Water Appropriations Subcommittee approved a measure that would maintain $134 million in prior year appropriations for FutureGen at Mattoon, IL.

July: Southern Illinois University's Clean Coal Review Board voted to award $2 million in grants for the FutureGen project for gasification, plant production and plant efficiency studies. The Alliance matched the grant and spent approximately $6 million on engineering and cost control studies.

December 12: The Alliance and Coles Together, a non-profit economic development organization in Coles County where FutureGen would be built, combined funds to purchase more than 420 acres of land in Mattoon, IL for approximately $7 million.

PHASE 3

2009

January 29: A bipartisan group of senators including Dick Durbin (D-IL), Kit Bond (R-MO), Claire McCaskill (D-MO) urged Secretary Chu to release the Record of Decision (ROD) certifying that a $1.8 billion coal-fueled experimental power plant would be built in Mattoon, IL.

February: A Government Accountability Office report showed that the DOE miscalculated the cost of FutureGen at $1.8 billion. GAO showed that in constant 2005 dollars from DOE's predicted cost estimate of $950 million, the Alliance's predicted cost for FutureGen increased by 37% or $370 million to $1.3 billion by 2017.

February 17: The American Recovery and Reinvestment Act provided $1.073 billion to the FutureGen program to advance construction of a plant built in Mattoon, IL.

June: Southern Company withdrew from the Alliance stating its intention to focus on coal gasification in its Kemper County, MS power plant and a carbon research center in Wilsonville, Alabama.

June 12: The Alliance and DOE reached an agreement to proceed with the preliminary design and cost estimate of FutureGen, estimated at $2.4 billion.

July 1: American Electric Power pulled out of the Alliance stating that FutureGen was moving too slowly and the company wanted to focus on carbon-sequestration projects like the Mountaineer plant in West Virginia.

July 14: The DOE issued the ROD, a final public decision that certifies that the Mattoon, IL site meets environmental requirements for the project.

September 2: The Alliance board of directors elected Steven Winberg of CONSOL Energy Inc. as the new chairman of the board to replace outgoing chairman Paul Thompson.

2010

January 12: The Illinois Finance Authority passed a resolution (Resolution Number 2010-01-09) providing the necessary financial mechanisms to issue bonds to help fund FutureGen.

January 30: Exelon Corporation, one of the nation's largest electric utilities, joined FutureGen Alliance.

February 8: Caterpillar Inc., a world-leading manufacturer of construction and mining equipment, joined the FutureGen Alliance.

PHASE 4

August 5: Secretary Chu announced the administration's new FutureGen 2.0 project, which would retrofit Ameren's existing power plant in Meredosia, IL with oxy-combustion technology at a 202 MW oil-fired unit. FutureGen 2.0 would be funded by $1 billion stimulus money and $247 million in private funds.

August 11: After DOE announced that Mattoon, IL would serve as the storage site for CO2 captured in Meredosia, IL, Coles Together removed Mattoon from participation in the FutureGen 2.0 project.

August 31: The Alliance Board of Directors offered support to DOE on its new FutureGen 2.0 program if mutual agreement on terms and conditions could be reached in fall, 2010.

September 28: DOE signed final cooperative agreements with the Alliance and Ameren Energy Resources that formally commit $1 billion in ARRA funds.

October 25: The Alliance issued requests for regions to submit proposals for hosting the carbon dioxide storage site.

December 20: Four Illinois counties (Christian County, Douglas County, Fayette County and Morgan County) were selected to advance to the next stage of site selection.

PHASE 4 CON'T

2011

February 28: The Alliance announced that it had selected Morgan County, IL as the location for the FutureGen 2.0 storage site.

April 5: The Alliance created FutureGen Citizens' Board to receive community feedback on geologic storage of CO2 and on visitors and training facilities.

June 7-9: DOE held public hearings in Taylorville, Tuscola and Jacksonville, IL about the environmental effects of storing carbon dioxide under these communities.

August 23: Illinois Legislature passed the "Carbon Dioxide Transportation" (S.B. 1821) legislation to aid with pipeline construction from Meredosia to the storage site in Morgan County.

October: The Alliance completed Phase 1: Pre-Front End Engineering Design (Pre-FEED) work which included plant design, estimated project cost and basis for applying for NEPA and other permits. The Pre-FEED report showed that the price of FutureGen increased from $1.3 billion to $1.65 billion.

October 4: Ameren said it will close its Meredosia and Hutsonville plants in Illinois because it could not afford to implement EPA regulations issued in July to reduce sulfur dioxide by 73% and nitrogen oxide by 54% from 2005 levels.

October 14: Developers began drilling a test well of about 5,000 feet in Springfield in Morgan County, IL. The plan is to pump carbon dioxide into the well in 2016.

October 31: Illinois Legislature passed "Clean Coal FutureGen for Illinois Act of 2011" (S.B. 2062) addressing liability management and "providing the FutureGen Alliance with adequate liability protection, land use rights, and permitting certainty to facilitate the siting of the FutureGen Project in Illinois."

November 28: The Alliance negotiated an option to purchase portions of the Meredosia Energy Center from Ameren in order to continue the development of the FutureGen 2.0 power plant.

December 20: The FutureGen Alliance successfully completed drilling a 5,000 foot characterization well in Morgan County, IL and preliminary data indicated that the local geology is suitable for carbon dioxide storage.

2012

January 26: A service rig was installed over the well in order to conduct hydrologic and geologic testing.

January 26: The Alliance finalized the sale of land originally purchased from Coles Together for the original FutureGen project. The conclusion of this transaction will allow Coles Together to continue to pursue economic development opportunities for the site and the Alliance will redirect funds from the sale to the FutureGen 2.0 project in Morgan County.

April: A revised Phase 1 report was submitted to the DOE with changes in project structure and schedule. The plant was rescaled to 168 MW capacity and the start of commercial operations changed from late 2015 to 2017.

July: After months of geologic and engineering studies, the Alliance confirmed Morgan County, IL as FutureGen 2.0's CO2 sequester site (the Alliance was considering Christian and Douglas Counties as alternative sequestration sites). The Alliance secured underground rights and began its application to the EPA for a CO2 storage permit.

December: The Illinois Commerce Commission voted 3-2 to approve a power procurement plan for the state that requires utilities to purchase all the electricity generated by the FutureGen 2.0 facility for 20 years. The order requires Commonwealth Edison and Ameren Illinois to purchase all of the gross electricity generated at the FutureGen 2.0 facility beginning in 2017.

2013

February: DOE approves the start of Phase 2 of the FutureGen 2.0 project.

February 28: The Illinois Competitive Energy Association, which represents alternative retail electric suppliers in the state, filed notice with the Illinois Appellate Court that it would challenge FutureGen's sourcing agreement, approved by the Illinois Commerce Commission (ICC) by a 3-2 vote in December. Commonwealth Edison (ComEd), one of the state's two main utilities, also announced its plans to challenge the sourcing agreement.

September 20: The U.S. Environmental Protection Agency (EPA) re-proposes a standard that would limit emissions of CO2 from new fossil fueled power plants. The re-proposed rule virtually requires that new coal-fired power plants install CCS to meet the new emissions standards.

October 25: DOE issues the Final Environmental Impact Statement (EIS) for the FutureGen 2.0 project.

2014

January 8: EPA publishes the re-proposed rule for limiting emissions from new power plants in the Federal Register, triggering a 60-day public comment period.

January 16: DOE issued a favorable Record of Decision (ROD) for FutureGen as part of the National Environmental Policy Act (NEPA) process.

2015

September 30: ARRA award expires

Sources: Information for the FutureGen Timeline has been acquired from the following sources.

Articles from the St. Louis Business Journal between April 22, 2003 and December 7, 2011 by various authors, http://www.bizjournals.com/stlouis/search/results?q=FutureGen.

Christa Marshall, "FutureGen Carbon Capture Project Affirms Main Storage Site," Environment & Energy, July 18, 2012, http://www.eenews.net/climatewire/print/2012/07/18/6.

D.K. McDonald, M. Estopinal, and H. Mualim, "FutureGen 2.0: Where Are We Now?", (Technical Paper, Babcock & Wilcox Power Generation Group, Inc., 2012), http://www.babcock.com/library/pdf/BR-1870.pdf.

Heartland Coalfield Alliance, "FutureGen 2.0 June Scoping Meetings," press release, June 2, 2011, http://heartlandcoalfieldalliance.org/futuregen-2-0-scoping-june-meetings/.

John Reynolds, "Ameren power plant closures: Fewer jobs, cleaner air.," The State Journal Register, October 4, 2011, http://www.sj-r.com/breaking/x432920643/Ameren-cites-EPA-rules-in-closure-of-2-Illinois-plants?zc_p=0.

Katherine Ling, "Senate panel freezes funding for restructured FutureGen," Environment & Energy, July 8, 2008, http://www.eenews.net/eenewspm/climate_change/2008/07/08/3.

Mark Chediak and Katarzyna Klimasinska, "AEP, Southern Withdraw From FutureGen Coal Project (Update2),"

Bloomberg.com, June 24, 2009, http://www.bloomberg.com/apps/news?pid=newsarchive&sid=aBeVHVGtr7KE.

Press Releases from The FutureGen Alliance between September 13, 2005 and July 17, 2012, http:/ /www.futuregenalliance.org/news/press-releases/.

Russell Gold, "Taking Lumps: Futuregen Backers Back Out.," Wall Street Journal, June 5, 2009, http://blogs.wsj.com/environmentalcapital/2009/06/25/taking-lumps-futuregen-backers-back-out/.

Tamar Hallerman, "Ill. Regulators Approve 20-Year Power Contract for FutureGen," GHG Reduction Technologies Monitor, December 21, 2012, http://ghgnews.com/index.cfm/ill-regulators-approve-20-year-power-contract-forfuturegen/.

The text of the Illinois Finance Authority Board Meeting on January 12, 2010 regarding," Resolution Number 2010-01-09 "A Resolution in Support of the Non-Profit Clean Coal FutureGen Project in Mattoon, Illinois" can be viewed at http://www.il-fa.com/public/boardbooks/media-1-12-10.pdf.

The text of Illinois S.B. 1821 can be viewed at http://www.ilga.gov/legislation/publicacts/fulltext.asp?Name=097- 0534.

The text of Illinois S.B. 2062 can be viewed at http://www.ilga.gov/legislation/publicacts/97/097-0618.htm.

U.S. Congress, House Committee on Science and Technology, Subcommittee on Energy and Environment, FutureGen and the Department of Energy's Advanced Coal Programs, 111th Cong., 1st sess., March 11, 2009.

U.S. Congress, Senate Committee on Appropriations, Subcommittee on Energy and Water Development, Department of Energy's Decision to Restructure the FutureGen Program, 110th Cong., 2nd sess., May 8, 2008, 110-826.

U.S. Department of Energy Office of Fossil Energy, FutureGen Integrated Hydrogen, Electric Power Production and Carbon Sequestration Research Initiative: Energy Independence through Carbon Sequestration and Hydrogen from Coal, Report to Congress, March 4, 2004, http://www.netl.doe.gov/technologies/coalpower/fuelcells/publications/fuelcell/fc-cleanup/futuregen_report_march_04.pdf.

U.S. Department of Energy Fossil Energy Techline, "Department of Energy Formally Commits $1 Billion in Recovery Act Funding to FutureGen 2.0," press release, September 28, 2010, http://www.fossil.energy.gov/news/ techlines/2010/10048-DOE_Formally_Commits_%241_Billion_to.html.

U.S. Government Accountability Office, Clean Coal DOE's Decision to Restructure FutureGen Should Be Based on a Comprehensive Analysis of Costs, Benefits, and Risks, GAO-09-248, February 13, 2009.

End Notes

[1] Congress first appropriated funds specifically for FutureGen in FY2005.

[2] U.S. Department of Energy National Energy Technology Laboratory, "Secretary Chu Announces FutureGen 2.0: Awards $1 Billion in Recovery Act Funding for Carbon Capture and Storage Network in Illinois," press release, August 5, 2010, http://www.netl.doe.gov/publications/ press/2010/10033-Secretary_Chu_Announces_FutureGen_.html.

[3] Email correspondence with Jeff Hoffman of the Office of Major Demonstrations in the Department of Energy's National Energy Technology Laboratory.

[4] Ibid.

[5] FutureGen Alliance, "FutureGen 2.0," press release, February 24, 2011, http://www.future genalliance.org/pdf/ FutureGenFacts.pdf.

[6] University of Illinois, Institute of Government and Public Affairs, Regional Economics Applications Laboratory, Economic Impacts of FutureGen 2.0 on Illinois and Local Economies, Urbana, IL, June 2013, http://www.jredc.org/ resources/Economic_Impact_FutureGen2%200_Hewings_6-2013_Final.pdf.

[7] For a more detailed examination of the science of climate change, see CRS Report RL34513, Climate Change: Current Issues and Policy Tools, by Jane A. Leggetthttp://www.crs.gov/pages/ Reports.aspx?PRODCODE=RL34513.

[8] For a more detailed examination of DOE's CCS program, see CRS Report R42496, Carbon Capture and Sequestration: Research, Development, and Demonstration at the U.S. Department of Energy, by Peter Folger.

[9] D. K. McDonald, M. Estopinal, and H. Mualim, "FutureGen 2.0: Where Are We Now?," (Technical Paper, Babcock & Wilcox Power Generation Group, Inc., 2012), pp. 2-3, http://www.babcock. com/library/pdf/BR-1870.pdf. (Hereinafter referred to as McDonald.) Babcock & Wilcox Power Generation Group is a technology provider for FutureGen 2.0 carbon capture project.

[10] McDonald et al., 2012, p. 4.

[11] Tennille Tracy, "Ameren Quits FutureGen Pollution Project," The Wall Street Journal, November 28, 2011.

[12] "At the Major CCS Projects: HECA, FutureGen," GHG Reduction Technologies Monitor, July 20, 2012, http://ghgnews.com/index.cfm/at-the-major-ccs-projects-futuregen-20-heca/?mobile Format=false. (Hereinafter referred to as GHG ReductionTechnologies Monitor, July 20, 2012.)

[13] Tamar Hallerman, "Ill. Regulators Approve 20-Year Power Contract for FutureGen," GHG Reduction Technologies Monitor, December 21, 2012, http://ghgnews.com/index.cfm/ill-regulators-approve-20-year-power-contract-forfuturegen/. (Hereinafter referred to as GHG Reduction Technologies Monitor, December 21, 2012.)

[14] GHG Reduction Technologies Monitor, December 21, 2012.

[15] See FutureGen 2.0 Industrial Alliance, Community Corner Archive, http://www.futuregenal liance.org/communitycorner/2013/03/.

[16] Ken Humphreys, CEO of the FutureGen 2.0 Industrial Alliance, personal communication, April 18, 2013.

[17] DOE, EIS-0460: Final Environmental Impact Statement, FutureGen 2.0 Project, Morgan County, Illinois, October 25, 2013, http://energy.gov/nepa/downloads/eis-0460-final-environmental-impact-statement.

[18] Environmental Protection Agency, "EIS No. 20130314, Final EIS, DOE, IL, FutureGen 2.0, Project, Review," 78 Federal Register 65643, November 30, 2013.

[19] Christa Marshall, "FutureGen Gets Final Nod from DOE," ClimateWire, January 17, 2014, which cites a statement by FutureGen Industrial Alliance CEO Ken Humphreys.

[20] The permit would be issued pursuant to the Safe Drinking Water Act, Underground Injection Control Program at EPA. The FutureGen Industrial Alliance has submitted applications for four Class VI CO2 sequestration wells. See http://www.epa.gov/r5water/uic/futuregen/.

[21] DOE partnered with Ameren to retrofit the corporation's obsolete 200 MW power plant in Meredosia, IL, with oxycombustion technology. The plans are for the retrofitted power plant to capture 90% of emitted carbon dioxide and transport it from Meredosia to a storage site in Morgan County, IL, to store up 1.3 million tons of carbon dioxide per year. The portion of funding from ARRA is $589,744,000. After Ameren withdrew from the cooperative agreement, the FutureGen Industrial Alliance took responsibility for the capture technology portion of the project as well as the pipeline and sequestration portion.

[22] U.S. Department of Energy National Energy Technology Laboratory, FutureGen 2.0, Project Facts, June 2011, http://www.netl.doe.gov. Funds apportioned from the DOE to the FutureGen Alliance include $404,985,000 from ARRA funds and $53.6 million from prior year appropriations toward the FutureGen project through the Office of Fossil Energy.

[23] FutureGen 2.0 Alliance, "FutureGen 2.0, Frequently Asked Questions—General," December 2013, http://www.futuregenalliance.org/wp-content/uploads/2013/12/FutureGen-FAQ-General-Dec-2013.pdf. According to the July 2013 DOE FutureGen 2.0 Fact Sheet, the total cost of the project is estimated to be $1,774,849,504, of which $1,048,348,11 would be covered by DOE, and the remaining portion would be covered by the FutureGen Industrial Alliance. See http://www.netl.doe.gov/publications/factsheets/project/FE0001882-FE0005054.pdf. Some more recent reports put the project cost at $1.68 billion, see Christa Marshall, "FutureGen Gets Final Nod from DOE," ClimateWire, January 17, 2014.

[24] McDonald et al., 2012, p. 4.

[25] Recovery.gov website, FutureGen Industrial Alliance, Inc., October 2013, http://www.recovery.gov/arra/ Transparency/RecoveryData/Pages/Recipient.aspx?duns=603703799.

[26] Tamar Hallerman, "DOE, FutureGen Eye Tight Project Timeline," GHG Reduction Technologies Monitor, October 25, 2013, http://ghgnews.com/index.cfm/vol-8-issue-46/doe-futuregen-alliance-eye-tight-project-timeline/.

[27] Ibid.

[28] Massachusetts Institute of Technology, The Future of Coal: An Interdisciplinary MIT Study (2007), p. xiii.

[29] Philip Webre and Samuel Wice, Federal Efforts to Reduce Cost of Capturing and Storing Carbon Dioxide, Congressional Budget Office, June 2012, p.5, http://www.cbo.gov/sites/default/files/cbofiles/attachments/43357-06- 28CarbonCapture.pdf.

[30] Ameren Energy Resources Company, LLC, "Two Ameren Merchant Generating Company Energy Centers to Cease Operations," press release, October 4, 2011, http://ameren.mediaroom.com/index.php?s=43&item=981. In January 2013 Ameren Energy Resources agreed to sell a portion of the Meredosia Energy Center to the FutureGen 2.0 Alliance for the FutureGen 2.0 project. Ownership is expected to formally transfer to the FutureGen 2.0 Alliance in 2014 prior to the start of construction of the and if all contractual conditions are met. See http://www.futuregenalliance.org/ community-corner/2013/03/.

[31] See CRS Report R42496, Carbon Capture and Sequestration: Research, Development, and Demonstration at the U.S. Department of Energy, by Peter Folger.

[32] Steve Koonin, DOE Strategic Plan, U.S. Department of Energy, December 8, 2010, p. 8, http://efcog.org/library/ council_meeting/SAMtg.120810/Presentations/Koonin,%20Steve.pdf.

[33] The proposal and background information is available at http://www2.epa.gov/carbon-pollution-standards/2013- proposed-carbon-pollution-standard-new-power-plants.

[34] Environmental Protection Agency, "Standards of Performance for Greenhouse Gas Emissions From New Stationary Sources: Electric Utility Generating Units," 79 Federal Register 1429, January 8, 2014.

[35] U.S. Congress, House Committee on Energy and Commerce, Subcommittee on Energy and Power, The American Energy Initiative: A Focus on EPA's Greenhouse Gas Regulations, 113th Cong., 1st sess., June 19, 2012, testimony of Steven E. Winberg, Vice-President, Research and Development, CONSOL Energy, Inc., p. 6.

[36] For a detailed discussion of the EPA's regulation of coal, see CRS Report R41914, EPA's Regulation of Coal-Fired Power: Is a "Train Wreck" Coming?, by James E. McCarthy and Claudia Copeland.

[37] The projects cited in the re-proposed rule are the Southern Company Kemper County Energy Facility, the SaskPower Boundary Dam CCS project, the Summit Power Texas Clean Energy Project, and the Hydrogen Energy California Project. The Boundary Dam project is a Canadian venture; the other three projects are in the United States and are receiving funding from DOE. See CRS Report R42496, Carbon Capture and Sequestration: Research, Development, and Demonstration at the U.S. Department of Energy, by Peter Folger for more information on DOE funding for CCS.

[38] For a detailed discussion of how natural gas is affecting electric power generation, see CRS Report R42814, Natural Gas in the U.S. Economy: Opportunities for Growth, by Robert Pirog and Michael Ratner.

[39] The other three are feasibility, costs, and size of emission reductions.

[40] See for example, the November 15, 2013, letter to EPA Administrator Gina McCarthy from Rep. Fred Upton, chair of the House Committee on Energy and Commerce, http://www.eenews.net/ assets/2013/11/22/document_daily_03.pdf; and the December 19, 2013, letter to Administrator McCarthy from Rep. Lamar Smith, chair of the House Committee on Science, Space, and Technology, http://science.house.gov/sites/republicans.science.house.gov/files/documents/ Letters/121913_mccarthy.pdf.

[41] Philip Webre and Samuel Wice, Federal Efforts to Reduce Cost of Capturing and Storing Carbon Dioxide, Congressional Budget Office, June 2012, pp. 14-15, http://www.cbo.gov/sites/ default/files/cbofiles/attachments/43357- 06-28CarbonCapture.pdf.

[42] Several CRS reports cover the issues of technology and cost of capturing CO2, as well as the challenge of storage capacity in the United States for captured CO2, regulatory challenges, public acceptance, and others. See CRS Report R41325, Carbon Capture: A Technology Assessment, by Peter Folger; CRS Report R42532, Carbon Capture and Sequestration (CCS): A Primer, by Peter Folger; CRS Report RL34601, Community Acceptance of Carbon Capture and Sequestration Infrastructure: Siting Challenges, by Paul W. Parfomak, and others.

INDEX

A

B

C

D

E

M

N

O

P